Memristor - An Emerging Device for Post-Moore's Computing and Applications

Edited by Yao-Feng Chang

Published in London, United Kingdom

IntechOpen

Supporting open minds since 2005

Memristor - An Emerging Device for Post-Moore's Computing and Applications
http://dx.doi.org/10.5772/intechopen.92517
Edited by Yao-Feng Chang

Contributors
Sergey G. Grigori Rusakov, Vladimir V. Rakitin, Xiaohan Wu, Ruijing Ge, Deji Akinwande, Jack C. Lee, Firman Simanjuntak, Sridhar Chandrasekaran, Debashis Panda, Aftab Salem, Themis Prodromakis, Mangal Das, Sandeep Kumar, Ai-Dong Li, Chang Liu, Lin Zhu, Lai-Guo Wang, Amirali Amirsoleimani, Dominique Drouin, Roman Genov, Fabien Alibart, Tony Liu, Serge Eccofey, Yao-Feng Chang, Ronald Tetzlaff, Alon Ascoli, Ahmet Şamil Demirkol, Ioannis Messaris, Majid Ahmadi, Mehri Teimoori, Arash Ahmadi

First published in London, United Kingdom, 2021 by IntechOpen
IntechOpen is the global imprint of INTECHOPEN LIMITED, registered in England and Wales, registration number: 11086078, 5 Princes Gate Court, London, SW7 2QJ, United Kingdom
Printed in Croatia

British Library Cataloguing-in-Publication Data
A catalogue record for this book is available from the British Library

Additional hard and PDF copies can be obtained from orders@intechopen.com

Memristor - An Emerging Device for Post-Moore's Computing and Applications
Edited by Yao-Feng Chang
p. cm.
Print ISBN 978-1-83968-956-7
Online ISBN 978-1-83968-957-4
eBook (PDF) ISBN 978-1-83968-958-1

For EU product safety concerns:
IN TECH d.o.o., Prolaz Marije Krucifikse Kozulić 3, 51000 Rijeka, Croatia, info@intechopen.com or visit our website at intechopen.com.

Meet the editor

Dr. Yao-Feng Chang is a memory engineer at Intel. He received a BS from National Sun Yat-sen University, Taiwan, in 2007; an MS in Electronics Engineering from National Chiao Tung University, Taiwan, in 2009; and a Ph.D. in Electrical Engineering from the University of Texas at Austin, USA, in 2015. He has more than 100 journal publications to his credit. His research interests include SiOx- and SiNx- Based ReRAM fabrication, characterization, and mechanism; stateful Boolean logic and synaptic post-Moore's computing and bio-inspiration biomimetic systems in electronic devices; and selector integration for large-scale, low-power array design.

Contents

Preface

The potential of machine learning and novel computing architecture can be exploited in the immediate future if more efficient hardware and devices are developed that meet the special requirements of bio-inspired computing or unconventional computing schemes. In this area, non-volatile memory technology using memristive devices (not restrained to any type of device) in combination with neuromorphic systems and memcomputing (memristor + computing) is a promising way to achieve such hardware. This book provides a platform for interdisciplinary research into unconventional computing with emerging physical substrates. It includes studies in areas such as biological modeling, mat #research erials physics and analytics, memristive devices in miniature scale, neuromorphic circuits, memcomputing, advanced arithmetic operations for logic applications, and other novel computing concepts and circuit schemes for potential biomimetic smart systems.

Yao-Feng Chang
Intel,
Flagstaff, AZ, USA

Section 1

Memristor - Device and Characterizations

Chapter 1

Memristors Based on 2D Monolayer Materials

Xiaohan Wu, Ruijing Ge, Deji Akinwande and Jack C. Lee

Abstract

2D materials have been widely used in various applications due to their remarkable and distinct electronic, optical, mechanical and thermal properties. Memristive effect has been found in several 2D systems. This chapter focuses on the memristors based on 2D materials, e. g. monolayer transition metal dichalcogenides (TMDs) and hexagonal boron nitride (h-BN), as the active layer in vertical MIM (metal–insulator–metal) configuration. Resistive switching behavior under normal DC and pulse waveforms, and current-sweep and constant stress testing methods have been investigated. Unlike the filament model in conventional bulk oxide-based memristors, a new switching mechanism has been proposed with the assistance of metal ion diffusion, featuring conductive-point random access memory (CPRAM) characteristics. The use of 2D material devices in applications such as flexible non-volatile memory (NVM) and emerging zero-power radio frequency (RF) switch will be discussed.

Keywords: Two-dimensional materials, transition metal dichalcogenide, non-volatile memory, resistive switching, atomristor

1. Introduction

Memristors have been studied for several decades and a large variety of materials has been utilized in memristors. One of the most representative and well-studied materials is metal oxide, which exhibits resistive switching phenomenon and has been widely used as the active layer in resistive random-access memory (RRAM). In the recent years, two dimensional materials have been discovered and developed rapidly as the most attractive novel materials. In 2004, the first two-dimensional material, graphene (consisting of a single layer of carbon atoms), was discovered by A. Geim and K. Novoselov [1]. Since then, the remarkable and diverse electronic, optical, mechanical and thermal properties have drawn much interest and inspired a large amount of 2D materials to be identified and analyzed, including transition metal dichalcogenides (TMDs), diatomic hexagonal boron nitride (h-BN), and monoatomic buckled crystals Xenes [2]. The 2D atomic sheets can be defined as atomically thin, layered crystalline solids, featuring intralayer covalent bonding and interlayer van der Waals bonding. These materials are recognized as two-dimensional since they represent the thinnest unsupported crystalline materials that can be realized. Graphene has been utilized in electronics devices mainly as the conductive electrodes since it is a zero-gap semiconductor. The material exhibits high electron mobility at room temperature with reported values of more than 15000 $cm^2\ V^{-1}\ s^{-1}$. Nevertheless, the zero-gap nature prevents its potential applications in field effect transistors (FET). MoS_2, a representative TMD material, has ~1.8 eV direct band gap in its monolayer form. Thus, it is suitable to be applied in FET devices [3]. In addition, hexagonal boron nitride (h-BN) also has drawn considerable attention among other 2D materials as a high band gap insulating material at ~5.9 eV, making it suitable for the production of ultrahigh mobility 2D heterostructures based on various types of 2D semiconductors [4]. Now the collection of 2D materials has been expanded to hundreds or expectedly thousands owing to more elemental and compound sheets uncovered [2, 5].

Non-volatile memory (NVM) has long been studied and developed by both academia and industry [6]. The most common non-volatile memory is flash memory [6, 7]. Although flash memory has advantages of fast read and write speed, low power consumption and less prone to damage compared with traditional hard disk drives, it has some drawbacks such as limited endurance and retention, high programming and erasing voltages, and the existing problems in small-area transistor structure like bias-temperature instability (BTI) or stress induced leakage current (SILC). In the search for the next-generation non-volatile memory, researchers have been working on various emerging alternatives, including ferroelectric random access memory (FeRAM), phase change memory (PCM), spin-transfer torque magnetic random access memory (STT-MRAM) and resistive random access memory (RRAM) [8, 9].

Among those emerging NVM, the RRAM devices show excellent endurance and retention compared with the commonly used flash memory, featuring lower power consumption, faster switching speed and better scalability [10]. The basic structure of a RRAM device is quite simple, basically a metal–insulator–metal (MIM) stacking. The conventional insulating material in RRAM is bulk metal oxides, such as SiO_2, TiO_2, or HfO_2 [11–14]. As the most common switching mechanism, conductive filaments will be formed in the insulator with external electrical bias. Depending on the formation and rupture of the conductive filament, the device can be repeatably switched between a high-resistance state (HRS) and a low-resistance state (LRS) and sustained without power supply. This is commonly referred to as the non-volatile resistance switching (NVRS) or memristive phenomenon. Recently, extensive works have been done in the development of RRAM devices not only in NVM application but also in brain-inspired neuromorphic computing due to its analog-like multi-state switching behavior [15–17].

In the past few years, motivated by the rapid development on 2D materials, researchers have found that several 2D materials also exhibit memristive phenomenon, expanding the NVRS materials to a large collection of ultrathin layered crystalline films. As a zero-gap 2D material, graphene is not suitable for resistance switching devices. On the other hand, graphene oxide has been successfully proved as the active layer in memristors [18]. MoS_2 is a representative 2D semiconductor, which has been found to show memristive effect in the form of 1 T phase [19]. In addition, Sangwan et al. reported that grain boundaries in monolayer MoS_2 film can produce NVRS in planar (horizontal) structure [20]. Nevertheless, the planar structure without 3D stacking ability has the limitation of low integration density. Another example is h-BN, a representative 2D insulator, which has been demonstrated to show the resistive switching behavior in multilayer nanosheets [21]. However, the monolayer 2D materials were not reported to exhibit the effect in vertical MIM configuration.

In this chapter, the memristors based on 2D monolayers (primarily TMDs and h-BN) are presented and discussed [22–25]. The devices (collectively labeled as atomristors) feature forming-free bipolar and unipolar switching, with relatively low switching voltages down to <1 V and large on/off current ratio of more than 10^6. Besides DC operation, the device can switch with fast switching speed by pulse operation (< 15 ns). An atomic-resolution Dissociation-Diffusion-Adsorption model has been proposed attributing the enhanced conductance to metal atoms/ions adsorption into intrinsic vacancies, a conductive-point mechanism supported by first-principle calculations and scanning tunneling microscopy (STM) characterizations [25, 26]. Besides voltage-sweep DC measurement, other characterization method like current sweeping and constant electric stress can be employed on the 2D-based memristors and illustrates more information in the resistive switching mechanisms [27, 28]. Benefit from the ultra-thin nature of the active layer, a novel application, RF switch, is realized based on the atomristors with operating frequencies covering the RF, 5G, and mm-wave bands and exhibits superior performance compared to those of existing solid-state switches [29–31]. The results discussed in this chapter have been organized and reproduced with permissions based on several representative publications in this field.

2. Fabrication of 2D-based memristors

A dozen 2D materials have been investigated for non-volatile resistive switching, including transition metal sulfides (MS_2, M = Mo, W, Re, Sn), transition metal selenides (MSe_2, M = Mo, W, Re, Sn, Pt), a transition metal telluride ($MoTe_2$), a TMD heterostructure (WS_2/MoS_2) and an insulator (h-BN). These selected 2D materials can be readily grown as mono or few layers with unambiguous characterization of material quality and thickness, using chemical vapor deposition (CVD) or metal–organic chemical vapor deposition (MOCVD) method [32, 33].

Two device structures were used for the 2D-based memory device fabrication. First is the typical crossbar device with the advantages of small-area capability and better probing condition. The schematic and optical image of MoS_2 crossbar device are shown in **Figure 1a** and **b**. Most of the electrical measurements were performed on the crossbar devices. The other structure, the litho-free and transfer-free device, was fabricated based on the 2D materials directly on metal foils to avoid possible residues or contamination induced by lithography or transfer process (schematic shown in **Figure 1c**). The crossbar device fabrication started with bottom electrodes (BE) patterning by electron beam lithography and 2 nm Cr/60 nm Au metal stack deposition on an SiO_2(285 nm)/Si substrate. Monolayer TMD was then transferred onto the fabricated substrate using a resist-free polydimethylsiloxane (PDMS) stamp transfer method. In this method, monolayer TMD was brought into conformal contact with PDMS. The substrate-TMD-PDMS system was subsequently soaked into diluted water. Since the original SiO_2 substrate is hydrophilic, it is easy for water to diffuse into the TMD-substrate interface, which helps separate the two layers. Then, the PDMS-TMD film was brought into contact with the target substrate with BE on it. The PDMS stamp was peeled off to leave monolayer TMD films on the target substrate. CVD h-BN was transferred onto BE from the Ni foil

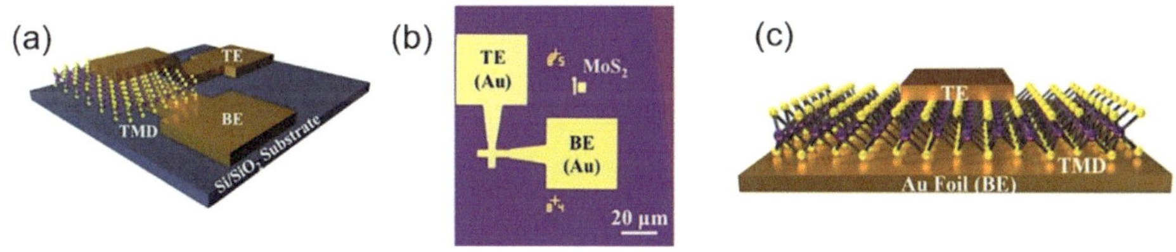

Figure 1.
(a, b) schematic and optical image of MIM structure of TMD crossbar device. (c) Schematic of TMD litho-free and transfer-free device based on MoS_2 grown on Au foil.

substrate using another poly(methyl methacrylate) (PMMA)-assisted wet transfer method. A thin layer of PMMA was spin coated onto the h-BN/Ni and then the Ni was etched away in 0.5 M ammonia persulfate solution. The PMMA/h-BN was rinsed in DI water to remove any etchant by-product before lifting by the target substrate with BE. The PMMA was then removed by immersing in acetone. For crossbar devices, top electrodes (TE) was patterned by e-beam lithography and deposited by e-beam evaporation using the same fabrication process as BE. In litho-free and transfer-free device, metal foils were used as global BE, and the TE (60 nm Au) was deposited via a shadow mask.

3. DC and pulse switching characteristics

DC electrical measurements were performed on as-fabricated devices consisting of atomic sheets with Au bottom and top electrodes and revealed memristive phenomenon in a dozen 2D systems (**Figure 2**). For instance, MoS_2, the prototypical

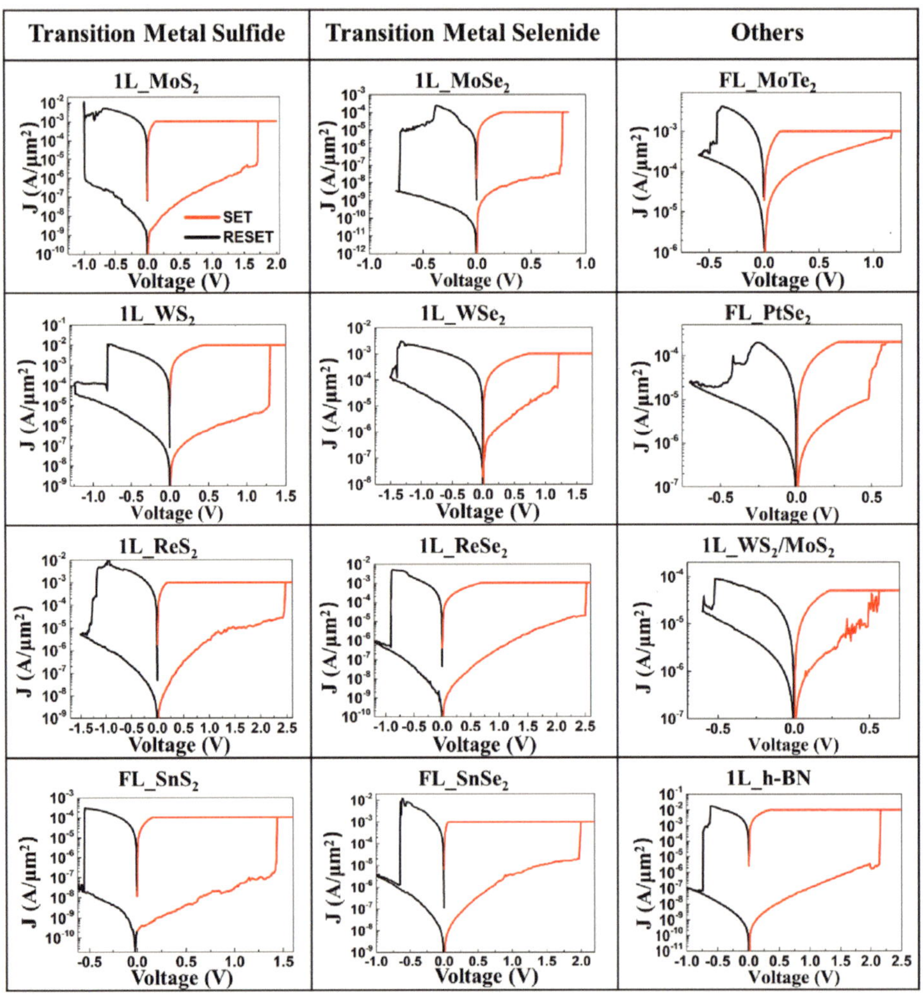

Figure 2.

Typical I-V curves of resistive switching behavior in crossbar devices for single-layer (1 L) MoS_2, WS_2, ReS_2, $MoSe_2$, WSe_2, $ReSe_2$, h-BN, and few-layer (FL) SnS_2, $SnSe_2$, $MoTe_2$, and litho-free device for monolayer WS_2/MoS_2 heterostructure, and multilayer $PtSe_2$. The y-axes are normalized as current density J.

TMD, featured low currents corresponding to a high-resistance state until the application of ~1.7 V, which "SET" the 2D-layer switch to a low-resistance state that maintains until a negative voltage is applied to "RESET" it. A compliance current is typically applied during SET process to prevent irreversible breakdown, while no compliance current is needed during RESET process. Interestingly, the monolayer non-volatile memory devices required no electro-forming step, a prerequisite in transition metal oxides (TMOs) that initializes a soft dielectric breakdown to form a conductive filament for following resistive switching operation [10]. Although some researches have shown that electroforming can be avoided by thickness scaling into the nm-regime, excessive leakage current from trap-assisted tunneling is a limiting consequence [10, 34]. Here, an ON/OFF ratio above 10^5 can be achieved in 2D NVRS devices, which highlights a defining advantage of crystalline monolayers over ultrathin amorphous oxides. These collective results of memristive phenomenon in representative atomic sheets allude to a universal effect in non-metallic 2D materials which opens a new avenue of scientific research on defects, charge, and interfacial phenomena at the atomic scale, and the associated materials design for diverse applications. Certain 2D memristors of the same MIM construction feature unipolar switching where voltage of the same polarity is used for both SET and RESET programming. Regarding the polarity dependence, the precise understanding of the factors that produce either bipolar or unipolar switching in 2D sheets is yet unclear and deserving of atomistic and unipolar switching is a complex competition among several parameters including lateral area, grain size, and modeling and microscopy studies for elucidation. A recent study in TMOs have suggested that the co-existence of bipolar compliance conditions, which may help the understanding of the phenomenon [35]. However, the underlying physics of unipolar switching has been previously established to be originated from electro-thermal heating that facilitates diffusion. A symptom of this effect is that a relatively higher RESET current is required to increase the local temperature to break the conductive link.

In most of the experiments, gold was selected as an inert electrode to rule out any switching effect that might arise from possible interfacial metal oxide formation. Furthermore, to rule out the undesirable contribution of polymer contamination from microfabrication, very clean devices including lithography-free and transfer-free devices (**Figure 3a**) were made, which also produced the memristive effect, alluding to an intrinsic origin. The lithography-free and transfer-free devices are based on monolayer MoS_2 grown directly on gold foil [36].

Previously reports have shown that line or grain boundary defects in polycrystalline 2D multi-layers play an intrinsic role in switching [37]. While it may be a possible factor in monolayers, it is not an exclusive factor as shown in **Figure 3b** from a vertical MIM device realized on a single-crystal CVD MoS_2, highlighting the potential role of localized effects. In addition, the NVRS phenomenon is not restricted to inert electrodes, since monolayer TMD with electrochemically active (Ag) electrodes can produce memristive effect as presented in **Figure 3c**. Moreover, monolayer graphene has also been demonstrated to be a suitable electrode option (**Figure 3d**).

Switching performance of retention time, DC switching cycling and variability was measured in 2D-based memristors. The NVRS devices present distinct advantages in terms of ultimate vertical scaling, down to an atomic layer thin with forming-free characteristics. By replacing metal electrodes with graphene, the entire memory cell can be scaled below 2 nm. Also, the transparency of graphene and the unique spectroscopic features of 2D materials provide the advantages of direct optical characterization for in-situ studies and in-line manufacturing testing. At an early stage, manual endurance data (**Figure 4a** and **b**) is not yet sufficient to meet the strict requirements for solid-state memory, a reflection of the nascent state of 2D atomristors compared to TMO memristors, which had similar endurance

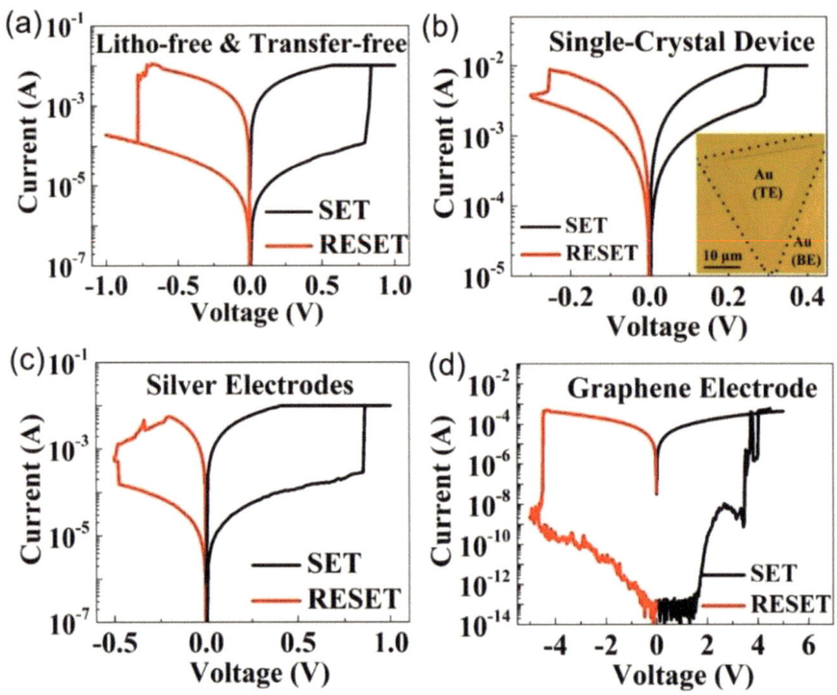

Figure 3.
Typical I-V curves of monolayer MoS_2 memristors with different device conditions, including (a) litho-free and transfer-free device, (b) single crystal device, (c) litho-free device with Ag as BE and TE, and (d) crossbar device with graphene as TE and Au as BE.

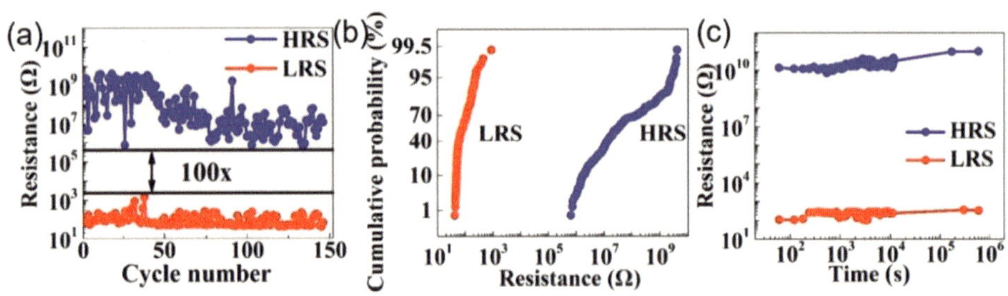

Figure 4.
(a,b) Endurance and resistance distribution of MoS_2 crossbar MIM device with 150 manual DC switching cycles. (c) Time dependent measurements of MoS_2 crossbar switch featuring stable retention over a week at room temperature.

($<10^3$ cycles) in early research but has now advanced above 10^6 cycles. Oxidation by interface engineering or doping may improve endurance performance, similar to what has been observed in amorphous-carbon memory devices [38]. Preliminary retention test of non-volatile states shows up to a week (**Figure 4c**), which is already sufficient for certain neuromorphic applications involving short and medium-term plasticity [39]. In addition, the sub-nanometer thinness of monolayers is promising for realizing ultra-high densities in 3D array architecture. As an estimation, at a loose pitch of 10 nm, an atomristor density of $10^{15}/mm^3$ would provide ample room to mimic the density of human synapses ($\sim 10^9/mm^3$). For single-bit single-level memory storage, it corresponds to a theoretical areal density of 6.4 $Tbit/in^2$.

Beyond DC characterization, pulse SET/RESET is feasible for 2D-based memristors (see **Figure 5** for monolayer h-BN device). The read I-V curves before

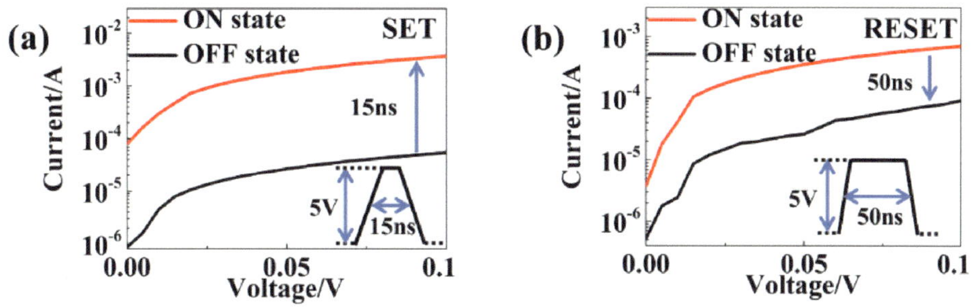

Figure 5.
(a) 15 ns SET and (b) 50 ns RESET pulse demonstration in h-BN memristor.

and after applying pulses clearly show the switching from OFF to ON state and from ON to OFF state, with 15 ns SET switching speed, and 50 ns RESET switching speed.

4. Parameter-dependent studies in 2D-based memristors

To gain more insights into the underlying mechanism(s), electrical measurements with the dependence of temperature, area scaling, compliance current, voltage sweep rate and layer thickness were performed with MoS_2 as the active layer owing to its greater material growth and characterization maturity. The low-voltage I-V characteristics at different temperatures are analyzed to explain the electron transport mechanisms at LRS and HRS. Metallic ohmic conduction can be deduced at LRS (**Figure 6a**) since the current decreases as the temperature increases, and the normalized conductance

$$G_n = (dI/dV)/(I/V) \tag{1}$$

is approximately one, a signature of linear transport that can be attributed to direct tunneling

$$J \propto KV \exp\left(\frac{-4\pi d\sqrt{2m^*\varphi}}{h}\right) \tag{2}$$

Where J is the current density, m^* is the effective mass, φ is the tunnel barrier height, h is Planck's constant, and K is proportional to the lateral area (A) and dependent on the barrier parameters (m, φ, d) [40]. d is the 2D barrier thickness. The direct tunneling model exhibits linear transport characteristics and is illustrated with an MIM band diagram (**Figure 6a**). Non-linear I-V characteristics are observed at HRS (**Figure 6b**), showing the current increasing as the temperature increases. The HRS data can be best fitted by the Schottky emission model with good agreement (**Figure 6c**) [40].

$$J \propto A^* T^2 \exp\left[\frac{-q\left(\phi_B - \sqrt{\frac{qE}{4\pi\varepsilon_r\varepsilon_0}}\right)}{kT}\right] \tag{3}$$

$$A^* = \frac{120m^*}{m_0} \tag{4}$$

where A* is the effective Richardson constant, m_0 is the free electron mass, T is the absolute temperature, q is the electronic charge, ϕ_B is the Schottky barrier height, E is the electric field across the dielectric, k is Boltzmann's constant, ε_0 is the permittivity in vacuum, and ε_r is the optical dielectric constant. For estimation, the effective thickness of ~1 nm is used and m^*/m_0 is ~1. The extracted barrier height is ~0.47 eV at 300 K, and the refractive index n is 6.84.

Area scaling studies have also been conducted and clearly show distinct profiles with the LRS relatively flat while the HRS has a more complicated relationship (**Figure 6d**). The LRS profile is consistent with the theory of a single (or few) localized filament(s) for TMO-based RRAM [10, 41]. With the area below 100 μm^2, the HRS resistance scales inversely with area owing to uniform conduction. For larger sizes, the resistance is relatively area-invariant, which can be attributed to the presence of localized grain boundaries. Note that the average domain size of typical CVD MoS_2 monolayer is ~10^2–10^3 μm^2. The current and resistance dependence on compliance current (see **Figure 6e** and **f**) reveal a linear relation that can be explained by an increase in the cross-sectional area of a single filament or to the formation of multiple filaments [41]. From the results of the temperature-dependent conduction experiments, the existence of Schottky barrier through TMD-metal interface from literatures [42, 43], and area-dependent studies, the NVRS behavior in MoS_2 devices can be explained by the proposed model

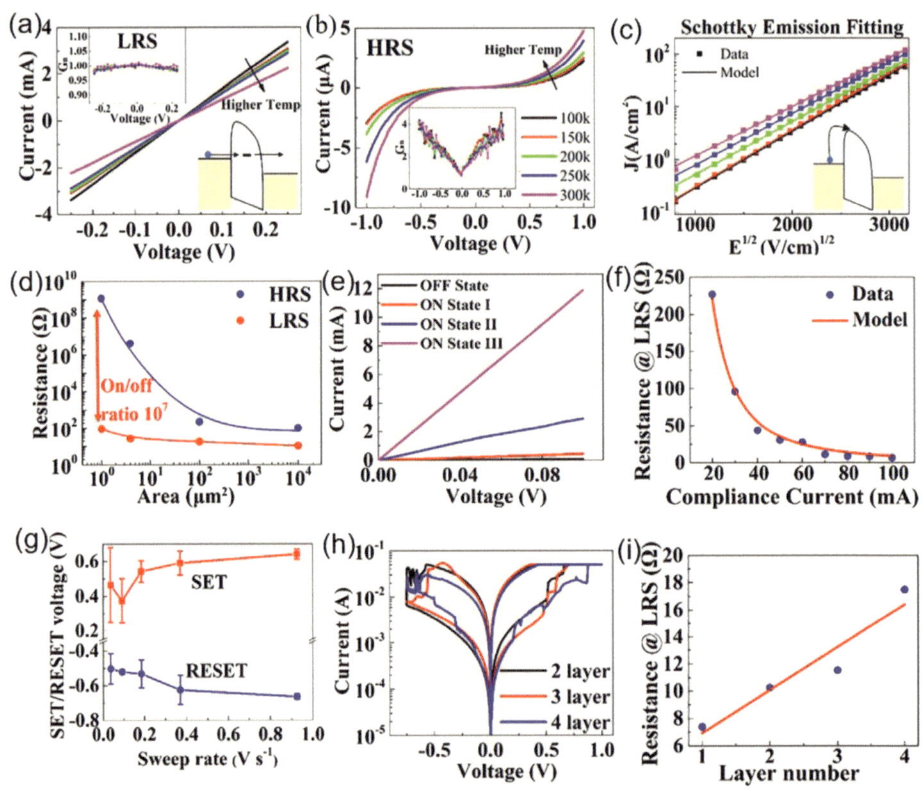

Figure 6.
Dependence of (a-c) temperature, (d) area scaling, (e, f) compliance current, (g) sweep rate, and (h, i) layer thickness of MoS_2 memristors.

that, during SET process, the electrons are transported through a filamentary-like 1D conductive link (or a virtual "conductive point"), and during RESET process, the conductive link is broken, resulting in a Schottky barrier at the device interfaces. Atomic level elucidation of the mechanisms(s) through advanced microscopy imaging and theoretical modeling is of great importance and is the focus of further research.

As to applications, the programmable resistance states are ideal for multilevel memory and neuromorphic computing. Moreover, the intrinsic low-resistance values ~5 Ω (**Figure 6f**), inspires a new application for low-power non-volatile RF switches. The dependence of the SET/RESET voltages on sweep rate (**Figure 6g**) suggests that more time is needed for ionic diffusion, which results in lower switching voltages. Layer dependent studies up to four layers demonstrate that the switching phenomena can be observed in few-layer 2D films (**Figure 6h**), with a distinction that the LRS resistance increases with layer number (**Figure 6i**).

5. Switching mechanisms in 2D-based memristors

To further elucidate the mechanism of NVRS phenomenon in 2D monolayers, a Dissociation-Diffusion-Adsorption (DDA) model has been proposed, (**Figure 7a**). In the vertical MIM structure, the symmetric electrodes choice (in most cases both TE and BE are gold) enables the formation of "conductive points" from either the top or bottom electrode. The first step is "Dissociation", which is based on the metal atom/ion dissociating from a cluster of metal atoms at the electrode-2D material interface. It is straightforward that this process depends on the choice of metal electrode. As discussed above, Au electrodes, as a noble metal, were selected to rule out potential effects from interfacial metal oxidation. It is also worth noting that Au has relatively low atomization enthalpy among various transition metals, thus can serve as an appropriate electrode [44]. First-principle calculation results have been performed and show that the dissociation energy required to move a Au atom sufficiently far from the bulk Au surface is 3.80 eV. For conventional conductive-bridge memory, the dissociation step is a common prerequisite that relies on the formation of metal ions to create a conductive filament and has been extensively investigated in previous reports, so the subsequent diffusion and adsorption steps will be the focus [45].

After Au atom/ion dissociates from the electrode, two scenarios may happen, with either directly adsorbing (chemical bonding) into a vacancy when they are close (Case 1), or it first weakly bonds to the pristine region and subsequently diffuses across the surface and finally finds a vacancy to fill and bond (Case 2). The two scenarios are illustrated in **Figure 7a**. Case 1 is a simpler scenario with only two steps "Dissociation" and "Adsorption" courtesy of the initial close position to a vacancy. On the other hand, Case 2 consists of all three steps and is expected to be more common since the adsorbed neutral Au atom (Au) or positively charged Au ion (Au^{+1}) in the pristine region are energetically favorable compared to their isolated states. Benefitting from the simplicity of Case 1, first-principle calculations for a collection of 12 materials were conducted, which have all been demonstrated to show NVRS behavior. In contrast, for the more probable Case 2, owing to the system complexity, only MoS_2 is analyzed as a prototypical monolayer in the TMD family.

In the simpler scenario Case 1, the dissociated Au is at first in an isolated state and tends to directly get adsorbed into the defect, resulting in the formation of conductive point that causes switching from HRS to LRS. It has been reported that the most common defects for 2D materials are vacancies, for example, S vacancy in MoS_2, Se vacancy in $MoSe_2$, B vacancy in BN, etc. The first-principle calculations

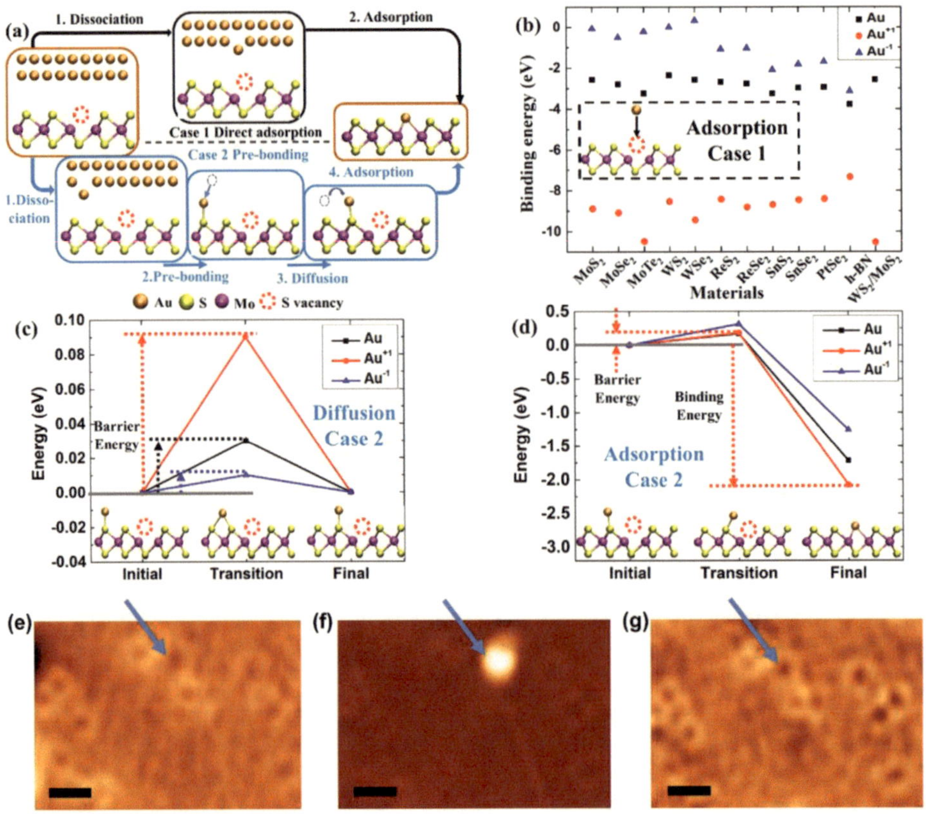

Figure 7.
(a-d) Calculated energy results and (e-g) STM observations for dissociation-diffusion-adsorption (DDA) model.

indicate that there is no barrier energy for Au to move in and bind with the defect site. This is straightforward to understand since isolated Au is unstable and the system energy tends to decrease as Au moves towards a defect site. In **Figure 7b**, the adsorption energy of Au atom/ion into a vacancy site has been calculated for various 2D materials. The negative adsorption energy (the energy difference between final state and initial state) means that adsorption is energetically favorable and releases energy, while a positive value means that the adsorption requires extra energy. Based on the calculations on diverse 2D materials, a common trend can be observed that both Au^{+1} and Au are energetically favorable to be adsorbed into defects, resulting in a SET process. To be more specific, Au^{+1} is the most favorable candidate, then neutral Au, and finally, negatively charged Au ion (Au^{-1}). A major reason for such a trend is that Au^{+1} is the most energetically unstable in its isolated vacuum state, thus releasing the most energy when covalently binding to a vacancy site, followed by the neutral Au atom and then Au^{-1}.

For the "Diffusion" step in Case 2, **Figure 7c** shows the calculated diffusion pathway and barrier energies (the energy difference between transition state and initial state) with Au moving along MoS_2 surface from the top of one S atom to the top of a neighboring S atom in the pristine region (without defects). Based on the first-principle calculations, the energy barrier for the Au atom/ion moving from one S atom site to another is quite low (< 0.1 eV), indicating that Au atom/ion can easily migrate around the pristine region at room temperature. This can be easily understood because the adsorption of Au atom/ion in the pristine region is weak, making them very mobile on the surface.

With regard to the final "Adsorption" step in Case 2, Au will diffuse to the atom close to the defect site, and eventually bind to it, since Au can easily move around the surface. **Figure 7d** shows the calculated energies for the transition and final states in the adsorption step. The low energy barrier (≤ 0.18 eV) indicates that Au/Au^{+1} can adsorb from the pristine region to the defect site, especially at high temperatures due to the Joule heating from the increased electrical current. In addition, this process can release a large amount of energy (≥ 1.72 eV). The low energy barrier and high energy released suggest that the adsorption of Au/Au^{+1} from the pristine region to the defect site is preferable both kinetically and energetically. However, the reversed process, for instance, the Au/Au^{+1} moving out from the vacancy site to the pristine region, has a much higher energy barrier (1.89 eV). Thus, it is much more difficult for Au/Au^{+1} to desorb from the vacancy site. As a result, Au/Au^{+1} can stably bind to the vacancy site, acting as a conducting point at LRS. During the RESET process, a high current usually passes through the conductive point, providing enough energy to overcome the barrier and driving Au/Au^{+1} away from the vacancy site. On the other hand, the Au^{-1} ion has the highest energy barrier and the smallest binding energy. As a result, the Au^{-1} ion is the least favorable to participate in the NVRS from both the kinetic and energetic viewpoints and it is not likely to play an essential role in resistive switching for both the scenarios discussed.

To provide experimental evidence to support the Dissociation-Diffusion-Adsorption model discussed above, STM measurement fitted with a gold tip was performed. STM was at first used for atomic resolution imaging of the MoS_2 surface to locate and identify the sulfur vacancies (**Figure 7e**). It was followed by a controlled physical contact of gold STM tip with the MoS_2 surface and voltage sweepings to emulate NVRS operation in a vertical MIM memory device. The STM image of the same location after SET shows a bright protrusion on the surface (**Figure 7f**). Stability of the site indicates it is not a diffusing atom. Instead, it is strongly bonded to the surface and identified as a gold atom absorbed into the sulfur vacancy [26]. RESET is realized by an opposite voltage sweeping where the gold atom is removed from the defect site (**Figure 7g**). The differences in sharpness and contrast of the STM images before and after the switching indicate that the tip apex has been changed due to the dissociation of a gold atom from the STM tip. In an extensive STM measurement, the STM tip was not only placed on top of the sulfur vacancy, but also in a pristine (defect-free) region. Compared with the I-V curves which resemble NVRS observed at the defect locations, electrical measurements on pristine regions reveal a tunneling-like I-V behavior with no switching phenomenon, suggesting the important role of defects (e.g. S vacancy) in a switching event [26].

6. Special operation methods of 2D-based memristors

To further investigate the NVRS phenomenon during SET process in the MoS_2 memristors, a current-sweep measurement method was introduced to the devices to get a more comprehensive understanding. **Figure 8a** shows the voltage–current (V-I) relationship by current-sweep method to SET a MoS_2 device. The transition starts at a HRS, followed by a gradual increase of both voltage and current. When the current reaches ~1.8 mA, the voltage suddenly decreases while the conduction current remained the same. In other words, the resistance of the device changes from a higher resistance state to a lower resistance state. Four subsequent voltage drops can be observed from 0.01 A to 0.03 A (as shown in the amplified figure). The device remains at the final lowest resistance state during and after the backward current sweeping, which indicates that a NVRS process from HRS to LRS

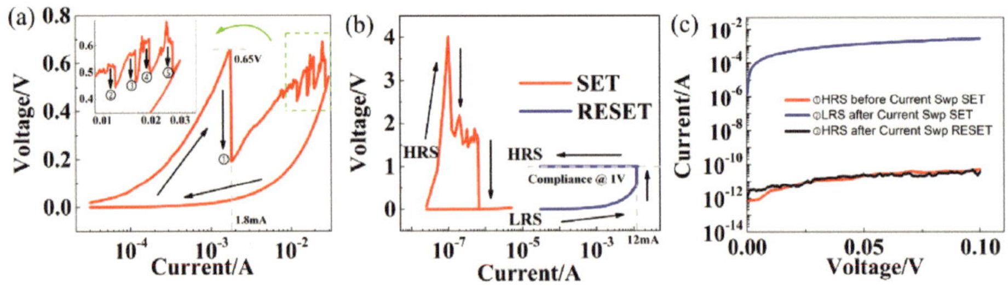

Figure 8.
(a, b) Current-sweep switching curves and (c) states reading behaviors in MoS_2 memristors.

(SET) is realized by current sweeping. Compared to the single-step SET process realized by voltage sweeping, multiple transition steps can be observed during current-sweep measurement. Note that the voltage for the first transition in current sweeping is ~0.65 V as shown in **Figure 8a**, which is very close to the SET voltage using voltage sweeping on the same device.

In **Figure 8b**, similarly, a multiple-step SET can be observed by current sweep. Moreover, the RESET process realized by current sweeping is presented in the same figure. When the current sweeps to ~12 mA, the voltage abruptly rises, suggesting a transition from LRS to HRS. This transition current is consistent with the RESET current (~10 mA) observed using voltage sweep method. Compared to RESET behavior by voltage sweeping, a compliance voltage is required in the case of current sweeping to avoid extremely high voltage across the device. Thus, it can be deduced that the RESET process is more likely to be a current/thermal-driven effect instead of voltage-driven effect. During RESET, a large amount of Joule heating can be induced by the high RESET current, which dissolves the conductive path first and then the Au ions will be migrated though porous regions or defects in the MoS_2 film, or back to the electrodes by reduction [10, 46]. This Joule heating effect is supported by the experimental observation that the transitions in voltage-sweep RESET (although sometimes with multiple steps) are sharp and sudden rather than gradual changes in 2D-based NVRS devices, a signature of Joule heating-dominated RESET process. Another evidence is that the MoS_2-based memristors can be switched in both bipolar and unipolar, which suggests it is not the electrical bias but the current level that plays a more important role in the RESET switching. **Figure 8c** shows the "READ" operations on the device before and after the current-sweep switching, which demonstrates the non-volatility of the NVRS behavior with a large on/off ratio of ~10^7. It can be observed that the resistance state after current-sweep RESET is consistent with the initial HRS state, which indicates the stable switching characteristics and alludes to a potential approach using current sweeping to improve the cycle-to-cycle variability at HRS, a long-standing issue for RRAM devices [47].

Figure 9a and **c** exhibit the switching curves of the SET process by current sweeping and RESET by voltage sweeping respectively on the same device. Similarly, the switching characteristics for current-sweep SET and voltage-sweep RESET tested on another device are shown in **Figure 9b** and **d**, respectively. Based on the statistical data of all the measured devices, a relationship can be established: normally for a device with single-step SET behavior by current sweeping (**Figure 9a**), the voltage-sweep RESET is also single-step (**Figure 9c**); on the other hand, for a device that has multiple steps during current-sweep SET (**Figure 9b**), a multiple-step RESET can be obtained by voltage sweeping (**Figure 9d**). With the

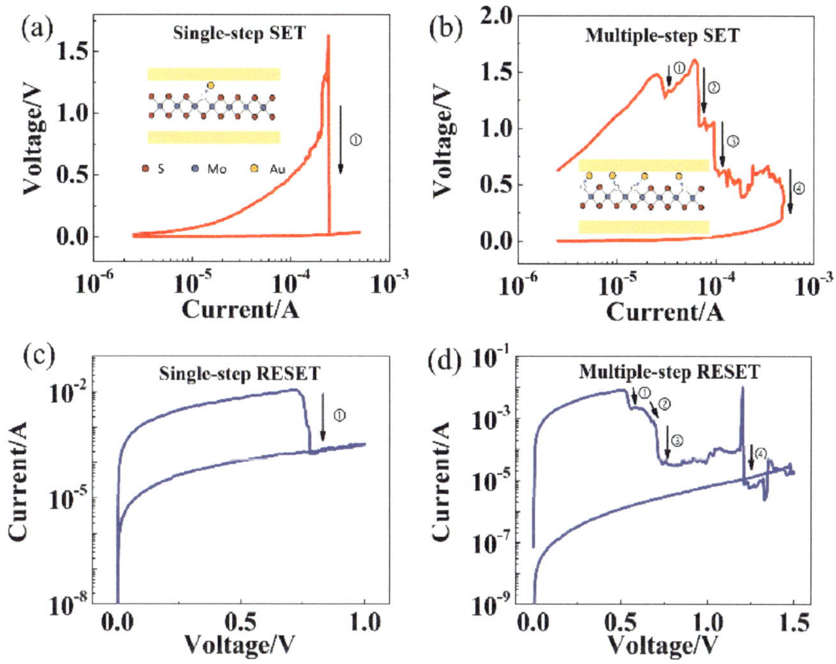

Figure 9.
The resistance switching characteristics of (a, b) current-sweep SET and (c, d) voltage-sweep RESET on the same device with similar transition behavior.

experimental results that show single or multiple transition steps, it can be inferred that multiple defect/vacancy-rich regions exist in the device area, which leads to single or multiple conductive points formation during NVRS.

Figure 10 presents the resistance evolution under constant voltage stress (CVS) on the devices at HRS. The working devices refer to the devices that exhibit stable switching characteristics and have been tested for several DC cycles and RESET to HRS before stress measurement. Then, relatively low constant voltage bias ($< V_{SET}$) is applied on the devices with positive CVS (**Figure 10a**) and with negative CVS (**Figure 10b**). It can be observed that the resistance changes from HRS to an even higher resistance state (labeled as HRS'). This phenomenon is opposite to the observation in the TMO-based devices, where the resistance is switched from HRS to LRS under CVS [48, 49]. Similar behavior has been observed with both positive and negative CVS, which can be related with the co-existence of unipolar and bipolar operations in MoS_2 memristors. Moreover, CVS test is performed on the fresh (as-fabricated) devices and shows similar HRS to HRS' transition (see **Figure 10c**). This phenomenon suggests that for both fresh devices and pre-RESET working devices, the commonly referred "HRS" is not the highest resistance, but actually an intermediate state that can still be modulated to a higher resistance state (HRS'). However, if the voltage stress goes higher than the SET voltage (**Figure 10d**), the device will switch to LRS and then fail due to high power.

The previously discussed DDA model with the assistance of metal atom/ion migration can be used to explain the NVRS phenomenon in MoS_2 memristors. The CVS test results provide more insights to this model with the tunable resistance states illustrated in **Figure 10e**. The existence of HRS' suggests that, a small portion of metal atoms may be embedded in the MoS_2 film at HRS, which could possibly

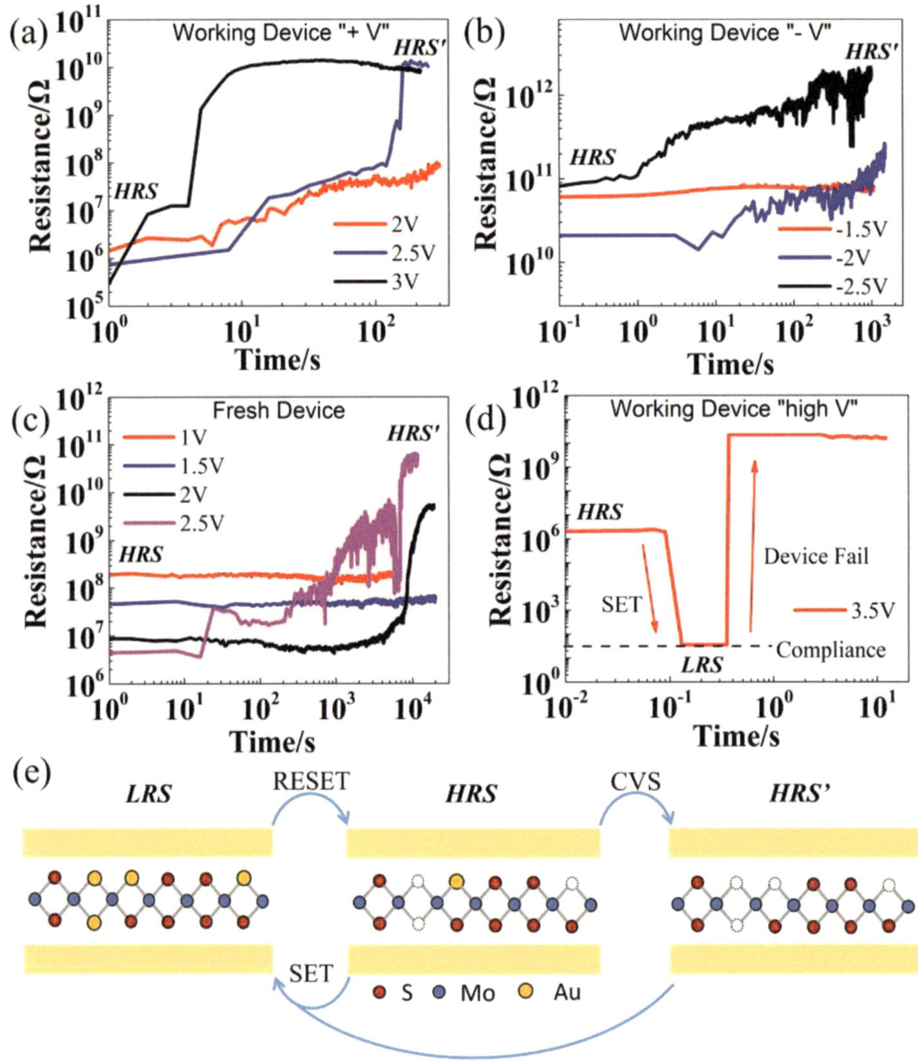

Figure 10.
(a-d) Resistance evolution under CVS MoS2-based memristors at HRS in different scenarios. (e) Illustration for the CVS process.

be induced by deposition process for as-fabricated devices or incomplete voltage-sweep RESET for working devices. Previous reports have shown experimental evidence to support this assumption that metal atoms can diffuse into the defects in 2D TMD films during the evaporation deposition process of TE confirmed by cross-sectional TEM images [50, 51]. These embedded metal atoms/ions are negligible in bulk metal oxides, but they can be important in the atomically thin MoS_2 sheets. With a relatively low voltage stress, these metal atoms tend to move out of the vacancies due to the accumulated Joule heating effect, which results in a transition to HRS'. This unique resistance evolution behavior under CVS suggests a distinct property for 2D materials. For traditional TMO-based bulk materials, the resistance state is typically controlled by the characteristics of the conductive filament and the "gap" region between the electrode and filament tip [10, 52]. While for 2D materials, the resistance state can be modulated by the interaction between atoms/ions from electrodes and interfacial vacancies, enabling atomic-level resistance control with advanced defect engineering for ultra-thin crystalline 2D materials.

7. Applications in flexible non-volatile memory and RF switch

Applications in flexible memory devices and RF switches were investigated based on 2D memristors. The high breaking strain and ease of integration of 2D materials on soft substrates can afford flexible non-volatile memory devices that can endure mechanical cycling (**Figure 11** for MoS_2-based memristor).

Non-volatile low-power RF switches represents another major application of atomristors. The low ON-state resistance values, below ~10 Ω, is critical for low-loss non-volatile RF switch circuits. The intrinsic experimental RF characteristics of monolayer MoS_2 switch show promising results of ~0.3 dB insertion loss in the ON-state (**Figure 12a**) and isolation below 20 dB in the OFF-state (**Figure 12b**) at frequencies up to 50 GHz [29]. By using monolayer h-BN as the active layer in the RF switch, the device exhibits a cutoff-frequency figure of merit of around 129 THz with a low insertion loss (≤0.5 dB) and high isolation (≥10 dB) from 0.1 to 200 GHz. In addition, it shows a high-power handling (around 20 dBm) and nano-second switching speeds, which are superior to those of existing solid-state switches [31]. This new application leads to the development of a nanoscale energy-efficient high-frequency solid-state switch technology for the rapidly growing communication systems in the 5G band and beyond.

A comparison between 2D atomristors discussed in this chapter and other representative 2D-based memory devices is presented in **Table 1**, highlighting the thinnest active layer thickness with superior switching properties and reliability as mentioned above.

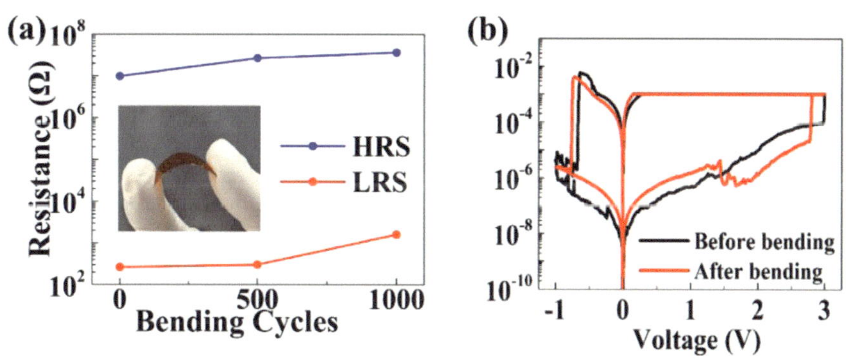

Figure 11.
(a) Stable resistance states and (b) switching behaviors after 1000 bending cycles at 1% strain.

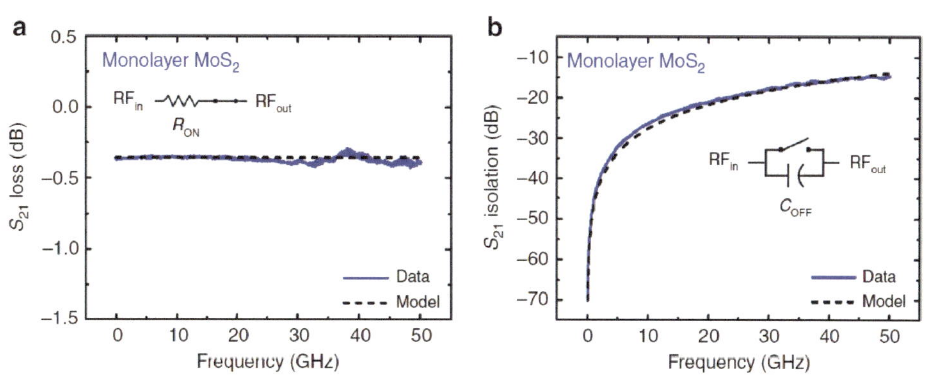

Figure 12.
Radio-frequency characterization of the MoS_2 RF switch: (a) insertion loss and (b) isolation.

Reference	**Active layer materials**	**Active layer thickness**	**Forming voltage**	**Switching voltage**	**Retention time (hrs)**	**Endurance**	**On/ Off ratio**	**Switching speed (ns)**
2D "Atomristors"	12 different 2D materials	0.3 nm–6 nm	Forming-free	0.5–4 V	>336	>240 DC cycles	$10 \sim 10^7$	15
Lee et al., Adv. Funct. Mater. 2020	MoS_2 in planar structure	~0.65 nm	Forming-free	~30 V	>24	>250 DC cycles	$10^2 \sim 10^3$	10^6
Xu et al., Nano Lett., 2019.	2 L MoS_2	~1.3 nm	Mostly Forming-free	~0.14–0.22 V	>5	>20 DC cycles	~10	N/R
Shi et al., Nat. Electron., 2018	5-7 L / 15-18 L h-BN	>1.5 nm	>1.8 V	0.1–6 V	>1	>100 DC cycles	$10 \sim 10^8$	2×10^4
Wang et al., Nat. Electron., 2018.	$MoS_{2-x}O_x$	~ 40 nm	~2 V	~1 V	>28	$>10^7$ pulse cycles	$\sim 10^2$	100
Zhao et al., Adv. Mater., 2017.	BNO_x	0.9–2.3 nm	Forming-free	0.6 V–1.7 V	>4	>100 DC cycles	$10^2 \sim 10^3$	10^6
Hao et al., Adv. Funct. Mater., 2016.	Degraded black phosphorus	~10 nm	N/R	1–2 V	>28	N/R	$\sim 3x\ 10^5$	N/R

Table 1.
Comparison of atomristors with other representative works in 2D memory.

8. Conclusion

In summary, a universal memristive phenomenon has been observed in 2D materials. These 2D-based memristors exhibit low switching voltage (<1 V), large on/off ratio (>10^6), fast switching speed (<20 ns), and forming-free characteristics. A mechanism based on metal atoms/ions adsorption into intrinsic vacancies producing an atomic-level conductive-point effect, has been proposed and supported by first-principle calculations and STM measurements. Constant voltage stress has been applied on the 2D-based memristors at high resistance state (HRS), revealing an additional higher resistance state that has not been discovered in conventional metal-oxide devices. Current sweeping method unveils the details hidden in the commonly used voltage-sweep curves, in which the transition step number could be attributed to the number of defects/vacancies. These open up a new materials space that might advance diverse applications including high-density neuromorphic computing, non-volatile memory fabrics, and zero-power RF switches.

Acknowledgements

This work was supported in part by the National Science Foundation (NSF) grant #1809017. The authors acknowledge use of Texas Nanofabrication Facilities supported by the NSF NNCI award #1542159.

Conflict of interest

The authors declare no conflict of interest.

Author details

Xiaohan Wu, Ruijing Ge, Deji Akinwande and Jack C. Lee*
Microelectronics Research Center, Department of Electrical and Computer Engineering, The University of Texas at Austin, Austin, Texas, USA

*Address all correspondence to: leejc@austin.utexas.edu

References

[1] Novoselov KS, Geim AK, Morozov SV, Jiang D, Zhang Y, Dubonos SV, et al. Electric field effect in atomically thin carbon films. Science. 2004;306(5696):666-669.

[2] Akinwande D, Petrone N, Hone J. Two-dimensional flexible nanoelectronics. Nature Communications. 2014;5:12.

[3] Mak KF, Lee C, Hone J, Shan J, Heinz TF. Atomically thin MoS2: A new direct-gap semiconductor. Physical Review Letters. 2010;105(13):4.

[4] Cassabois G, Valvin P, Gil B. Hexagonal boron nitride is an indirect bandgap semiconductor. Nature Photonics. 2016;10(4):262–+.

[5] Molle A, Goldberger J, Houssa M, Xu Y, Zhang SC, Akinwande D. Buckled two-dimensional Xene sheets. Nature Materials. 2017;16(2):163-169.

[6] Meena JS, Sze SM, Chand U, Tseng TY. Overview of emerging nonvolatile memory technologies. Nanoscale Research Letters. 2014;9:33.

[7] Zhao C, Zhao CZ, Taylor S, Chalker PR. Review on non-volatile memory with high-k dielectrics: Flash for generation beyond 32 nm. Materials. 2014;7(7):5117-5145.

[8] Upadhyay NK, Jiang H, Wang ZR, Asapu S, Xia QF, Yang JJ. Emerging memory devices for neuromorphic computing. Advanced Materials Technologies. 2019;4(4):13.

[9] Chen A. A review of emerging non-volatile memory (NVM) technologies and applications. Solid-State Electronics. 2016;125:25-38.

[10] Wong HSP, Lee HY, Yu SM, Chen YS, Wu Y, Chen PS, et al. Metal-Oxide RRAM. Proceedings of the Ieee. 2012;100(6):1951-1970.

[11] Zhou F, Chang YF, Byun K, Fowler B, Lee JC. Characterization of external resistance effect and performance optimization in unipolar-type SiOx-based resistive switching memory. Applied Physics Letters. 2014;105(13):4.

[12] Chen YC, Hu ST, Lin CY, Fowler B, Huang HC, Lin CC, et al. Graphite-based selectorless RRAM: Improvable intrinsic nonlinearity for array applications. Nanoscale. 2018;10(33):15608-15614.

[13] Chen YC, Lin CC, Hu ST, Lin CY, Fowler B, Lee J. A novel resistive switching identification method through relaxation characteristics for sneak-path-constrained Selectorless RRAM application. Scientific Reports. 2019;9:6.

[14] Bersuker G, Gilmer DC, Veksler D, Kirsch P, Vandelli L, Padovani A, et al. Metal oxide resistive memory switching mechanism based on conductive filament properties. Journal of Applied Physics. 2011;110(12):12.

[15] Burr GW, Shelby RM, Sebastian A, Kim S, Sidler S, Virwani K, et al. Neuromorphic computing using non-volatile memory. Advances in Physics-X. 2017;2(1):89-124.

[16] Hong XL, Loy DJ, Dananjaya PA, Tan FA, Ng C, Lew W. Oxide-based RRAM materials for neuromorphic computing. Journal of Materials Science. 2018;53(12):8720-8746.

[17] Ielmini D. Brain-inspired computing with resistive switching memory (RRAM): Devices, synapses and neural networks. Microelectronic Engineering. 2018;190:44-53.

[18] Pradhan SK, Xiao B, Mishra S, Killam A, Pradhan AK. Resistive switching behavior of reduced graphene

oxide memory cells for low power nonvolatile device application. Scientific Reports. 2016;6:9.

[19] Cheng PF, Sun K, Hu YH. Memristive behavior and ideal Memristor of 1T phase MoS2 Nanosheets. Nano Letters. 2016;16(1):572-576.

[20] Sangwan VK, Jariwala D, Kim IS, Chen KS, Marks TJ, Lauhon LJ, et al. Gate-tunable memristive phenomena mediated by grain boundaries in single-layer MoS2. Nature Nanotechnology. 2015;10(5):403-406.

[21] Pan CB, Ji YF, Xiao N, Hui F, Tang KC, Guo YZ, et al. Coexistence of grain-boundaries-assisted bipolar and threshold resistive switching in multilayer hexagonal boron nitride. Advanced Functional Materials. 2017;27(10):10.

[22] Wu XH, Ge RJ, Chen PA, Chou H, Zhang ZP, Zhang YF, et al. Thinnest nonvolatile memory based on monolayer h-BN. Advanced Materials. 2019;31(15):7.

[23] Ge R, Wu X, Kim M, Shi J, Sonde S, Tao L, et al. Atomristor: Nonvolatile resistance switching in atomic sheets of transition metal Dichalcogenides. Nano Letters. 2018;18(1):434-441.

[24] Ge RJ, Wu XH, Kim M, Chen PA, Shi JP, Choi J, et al. Atomristors: Memory Effect in Atomically-thin Sheets and Record RF Switches. 2018 Ieee International Electron Devices Meeting (Iedm). 2018:4.

[25] Ge R, Wu X, Liang L, Hus SM, Gu Y, Okogbue E, et al. A Library of Atomically Thin 2D Materials Featuring the Conductive-Point Resistive Switching Phenomenon. Advanced Materials. 2020;n/a(n/a):2007792.

[26] Hus SM, Ge RJ, Chen PA, Liang LB, Donnelly GE, Ko W, et al. Observation of single-defect memristor in an MoS2 atomic sheet. Nature Nanotechnology. 2020:6.

[27] Wu XH, Ge RJ, Akinwande D, Lee JC. Understanding of multiple resistance states by current sweeping in MoS2-based non-volatile memory devices. Nanotechnology. 2020;31(46).

[28] Wu XH, Ge RJ, Huang YF, Akinwande D, Lee JC. Resistance state evolution under constant electric stress on a MoS2 non-volatile resistive switching device. Rsc Advances. 2020;10(69):42249-42255.

[29] Kim M, Ge RJ, Wu XH, Lan X, Tice J, Lee JC, et al. Zero-static power radio-frequency switches based on MoS2 atomristors. Nature Communications. 2018;9:7.

[30] Kim M, Pallecchi E, Ge R, Wu X, Avramovic V, Okada E, et al., editors. Non-volatile RF and mm-wave Switches Based on Monolayer hBN. 2019 IEEE International Electron Devices Meeting (IEDM); 2019 7-11 Dec. 2019.

[31] Kim M, Pallecchi E, Ge RJ, Wu XH, Ducournau G, Lee JC, et al. Analogue switches made from boron nitride monolayers for application in 5G and terahertz communication systems. Nature Electronics. 2020;3(8):479-485.

[32] Chang HY, Yogeesh MN, Ghosh R, Rai A, Sanne A, Yang SX, et al. Large-area monolayer MoS2 for flexible low-power RF Nanoelectronics in the GHz regime. Advanced Materials. 2016;28(9):1818-1823.

[33] Ismach A, Chou H, Mende P, Dolocan A, Addou R, Aloni S, et al. Carbon-assisted chemical vapor deposition of hexagonal boron nitride. 2d Materials. 2017;4(2):10.

[34] Zhao L, Jiang Z, Chen HY, Sohn J, Okabe K, Magyari-Kope B, et al. Ultrathin (similar to 2nm) HfOx as the

Fundamental Resistive Switching Element: Thickness Scaling Limit, Stack Engineering and 3D Integration. 2014 Ieee International Electron Devices Meeting (Iedm). 2014:4.

[35] Yanagida T, Nagashima K, Oka K, Kanai M, Klamchuen A, Park BH, et al. Scaling effect on unipolar and bipolar resistive switching of metal oxides. Scientific Reports. 2013;3:6.

[36] Dumcenco D, Ovchinnikov D, Marinov K, Lazic P, Gibertini M, Marzari N, et al. Large-Area Epitaxial Mono layer MoS2. Acs Nano. 2015;9(4):4611-4620.

[37] Puglisi FM, Larcher L, Pan C, Xiao N, Shi Y, Hui F, et al. 2D h-BN based RRAM devices. 2016 Ieee International Electron Devices Meeting (Iedm). 2016:4.

[38] Santini CA, Sebastian A, Marchiori C, Jonnalagadda VP, Dellmann L, Koelmans WW, et al. Oxygenated amorphous carbon for resistive memory applications. Nature Communications. 2015;6:9.

[39] Indiveri G, Liu SC. Memory and information processing in neuromorphic systems. Proceedings of the Ieee. 2015;103(8):1379-1397.

[40] Chiu FC. A review on conduction mechanisms in dielectric films. Advances in Materials Science and Engineering. 2014;2014:18.

[41] Onofrio N, Guzman D, Strachan A. Atomic origin of ultrafast resistance switching in nanoscale electrometallization cells. Nature Materials. 2015;14(4):440-446.

[42] Kwon J, Lee JY, Yu YJ, Lee CH, Cui X, Honed J, et al. Thickness-dependent Schottky barrier height of MoS2 field-effect transistors. Nanoscale. 2017;9(18):6151-6157.

[43] Zhong HX, Quhe RG, Wang YY, Ni ZY, Ye M, Song ZG, et al. Interfacial properties of monolayer and bilayer MoS2 contacts with metals: Beyond the energy band calculations. Scientific Reports. 2016;6:16.

[44] Hilpert K, Gingerich KA. Atomization enthalpies of the molecules CU-3, AG-3, and AU-3. Berichte Der Bunsen-Gesellschaft-Physical Chemistry Chemical Physics. 1980;84(8):739-745.

[45] Peng CN, Wang CW, Chan TC, Chang WY, Wang YC, Tsai HW, et al. Resistive switching of Au/ZnO/Au resistive memory: An in situ observation of conductive bridge formation. Nanoscale Research Letters. 2012;7:6.

[46] Lanza M. A review on resistive switching in high-k dielectrics: A nanoscale point of view using conductive atomic force microscope. Materials. 2014;7(3):2155-2182.

[47] Sun W, Gao B, Chi MF, Xia QF, Yang JJ, Qian H, et al. Understanding memristive switching via in situ characterization and device modeling. Nature Communications. 2019;10:13.

[48] Wang GM, Long SB, Zhang MY, Li Y, Xu XX, Liu HT, et al. Operation methods of resistive random access memory. Science China-Technological Sciences. 2014;57(12):2295-2304.

[49] Lorenzi P, Rao R, Irrera F. Conductive filament evolution in HfO2 resistive RAM device during constant voltage stress. Microelectronics Reliability. 2015;55(9-10):1446-1449.

[50] Liu Y, Guo J, Zhu EB, Liao L, Lee SJ, Ding MN, et al. Approaching the Schottky-Mott limit in van der Waals metal-semiconductor junctions. Nature. 2018;557(7707):696−+.

[51] Kong LG, Zhang XD, Tao QY, Zhang ML, Dang WQ, Li ZW, et al.

Doping-free complementary WSe2 circuit via van der Waals metal integration. Nature Communications. 2020;11(1):7.

[52] Zahoor F, Zulkifli TZA, Khanday FA. Resistive random access memory (RRAM): An overview of materials, switching mechanism, performance, multilevel cell (mlc) storage, Modeling, and applications. Nanoscale Research Letters. 2020;15(1):26.

Chapter 2

Effect of Surface Variations on Resistive Switching

Mangal Das and Sandeep Kumar

Abstract

In this chapter, we study factors that dominate the interfacial resistive switching (RS) in memristive devices. We have also given the basic understanding of different type of RS devices which are predominantly interfacial in nature. In case of resistive random access memory (RRAM), the effect of surface properties on the bulk cannot be neglected as thickness of the film is generally below 100 nm. Surface properties are effected by redox reactions, interfacial layer formation, and presence of tunneling barrier. Surface morphology affects the band structure in the vicinity of interface, which in turn effects the movements of charge carriers. The effect of grain boundaries (GBs) and grain surfaces (GSs) on RS have also been discussed. The concentration of vacancies (O_v)/traps/defects is comparatively higher at GBs which leads to leakage current flow through the GBs predominantly. Such huge presence of charge carriers causes current flow through grain boundaries.

Keywords: resistive switching, interface, grain boundary, oxygen vacancy, surface morphologies

1. Introduction

In this chapter, the fundamentals of nanoionic redox based resistive switching materials are described including different modes of switching. The primary parameters have also been discussed which essentially affect the resistive switching behavior. In addition, this chapter encompasses the various physical as well as chemical phenomena occurring at nano level during resistive switching of devices and other related trending technological areas are also included. An introduction to three distinct kinds of redox based resistive switching materials is given followed by their short history and promising applications into device fabrication field. Further, a brief discussion related to requirement and optimization of device performance parameters has been incorporated along with future prospects, challenges and important industrial applications such as memory, logic circuits and so on. The elements that observe resistive switching are driven by a reversible phenomenon taking place between two terminals. This behavior depicts primarily two different resistance values of nonvolatile nature depending on the external electrical bias conditions [1]. However, such a reversible behavior obtained under continuous applied external stimuli can also be achieved within more than two resistance levels commonly called as multilevel resistive switching. The term 'nonvolatile' suggests the retention or preservation of change in resistance level after removal of the external stimuli. Such memristive materials are capable of memorizing these resistance values under the influence of stimuli [2]. Similar to phenomenon of resistive switching, there can be different possible switching scenarios such as magnetoresistive phenomenon i.e. spin-transfer torque, electrical effects like the leakage current *via* gate oxide layer containing trapped defects, change in structure/phase among amorphous and crystalline phases, and nanoionic redox phenomenon [1]. So, this chapter deals with the phenomenon of nanoionic redox realized due to ion movement within the two terminal based device structure elements and results into different resistance values of the material. In this chapter, a few terminology such as redox-based resistive switching random access memory (RRAM), resistive switching and/or memristive etc. is to be used frequently. Such terms have been essentially included to explain the fundamental physical characteristics of materials showing potential in multilevel switching or analog properties.

2. Resistive system and type of physical mechanism

Following the systematic and consistent study of RRAM devices, it has been realized that the resistive switching material (I) embedded between two metal electrodes (M) plays a key role in switching phenomenon in the metal–insulator–metal (MIM) stratified device structures. In general, MIM devices are referred to sandwiched stacks/layers of metals and insulator. In a simplistic way, the resistive switching effect can be identified phenomenology as the different switching occurring within stratified MIM layered structure. For such device structures, it is comprehensible to understand the different locations of switching while moving vertically from one metal electrode to other. Therefore, there can be following possible switching locations (**Figure 1**):

i. switching taking place in close proximity to interface of either metal electrode

ii. switching at or near middle of the electrode interfaces

iii. switching occurring over or along the whole path formed among interfaces of metal electrodes.

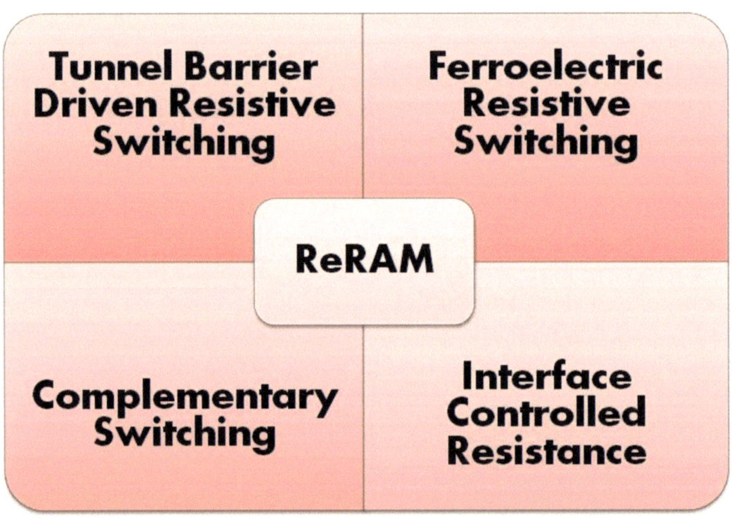

Figure 1.
Different types of resistive switching in which the formation of an interfacial layer between metal electrode and oxide plays a key role in device performance.

On the other hand, switching locations are also possible along perpendicular direction i.e. in the plane of device cross section (lateral direction). Due to this, following switching types are induced:

i. switching based on the formation and rupture of conducting filament (CF) called as filamentary switching.

ii. switching dependent upon contact area. In this case, the switching phenomenon participates to whole device cross section. This results into the scaling of current with the area of cross section. Usually, such a switching behavior is also termed as interface type switching due to the process taking place near the metal electrode.

3. Interface-type switching

The interface type switching is usually of uniform nature and shows area scalable characteristics. ReRAMs devices utilizing metal oxides (initially semiconductor or conductor) are frequently governed by interface type switching phenomena. Such a resistive switching behavior is primarily observed at the metal oxide and electrode interface. In this context, Baikalov *et al.* [1] investigate the resistance behavior of theperovskite oxide based memory devices. It has been demonstrated that an applied electric field significantly modifies the value of contact resistance measured among perovskite oxide and metal electrode. In order to understand the interface type resistive switching, many groups of researchers have given effort to establish different models for switching mechanism. These models are framed using ionic point defects (oxygen vacancies) and their drifted movement or electromigration within the metal oxide materials [1–5]. Some of the models are based on the formation of charge carrier i.e. electrons and/or holes trapping at defect sites [6] and observation of Mott transitions taking place at the interfaces [7, 8]. In the case of ReRAMs, low resistance value (in the range of kΩ) as well as the extended retention behavior (in the range of years) are realized in terms of models based on the modification of atomic or ionic configurations. Therefore, ionic and electronic mechanisms are used to examine the resistive switching characteristics and its origin at microscopic level. For ReRAMs, the interface type resistive switching is primarily predicted to dominant if the resistance of device scales with area. Therefore, Nb-doped $SrTiO_3$ is governed by interface type switching whereas filament based resistive switching prevails in NiO based memory cells [9, 10].

In case of valence change memory (VCM) devices, the resistive-switching phenomenon is understood by the chemical reactions and events taking place near or at the interface of metal electrode and oxide. While characterizing such VCM devices through spectroscopic techniques, the interface related switching behavior is investigated after averaging the area which essentially eliminates the need of any optical arrangements for focusing and/or magnification. The predefined interface dependent switching helps to designate the localized resistive switching phenomenon ascribed to CFs. Due to such interface i.e. thin oxide layer developed between insulator oxide and reactive metal electrode, the external stimuli driven electrochemical modification of interface considerably affects the performance of device. It is known that devices fabricated with reactive electrodes and insulating oxide materials show redox type reactions inducing movement of oxygen towards metal electrode (**Table 1**).

Switching Device Fabrication Technique of Switching layer	Retention (sec)	Endurance (cycle)	Switching Voltage	Type of Switching	Ref
Au/$LaMnO_3$/C-AFM Pulsed injection metal–organic chemical vapor deposition (PI-MOCVD)	—	—	V_{SET} = 4 V, V_{RESET} = −6 V,	Oxygen displacement, Mn oxidation state(+3.6 to +3.1), Work function decreases by 0.28 eV.	[11]
Ag/$Pr_{0.6}Ca_{0.4}MnO_3$/Al Pulsed laser deposition (PLD)	$3.6x10^5$	$12x10^3$	—	Formation and modulation of a rectifying interfacial AlO_x layer	[12]
Pt/$PbTiO_3$/ Nb:$SrTiO_3$(100) Hydrothermal method	$3x10^3$	$5x10^2$	V_{SET} = 4 V, V_{RESET} = −4, −6 V	Trap controlled spacecharge-Limited, M odulation of the Pt/ $PbTiO_3$Schottky-like junction	[13]
Pt/Nb:STO/Pt PLD	10^6	10^7	—	Pt/single crystal Nb:$SrTiO_3$ Schottky junction	[10]
Ti/PCMO/$SrRuO_3$ PLD	—	—	V_{SET} = 6.8 V, V_{RESET} = −6.8 V	Cacomposition, dependence on the RS characteristics, formation of amorphous TiO_ylayers at the interfaces	[14]

Table 1.
Comparison of different parameters for interfacial type RS devices.

3.1 Formation of a blocking layer on conducting oxides

The formation of an interface between oxide layer and an active electrode shows significant impact onto the different phenomena such as nonlinear transport of the charge carriers and affects performance of VCM type devices. In particular, the nonstoichiometric metal oxide layer demonstrates reversible oxidation and reduction at or near the interface which makes device functional. It is known that the interface type switching is mainly observed in conducting oxides like doped manganites utilizing metal electrodes e.g. Al, Ti and Ta etc. Such electrode metals exhibit relatively high oxygen affinity facilitating interface driven switching [9, 15, 16]. In case of $Pr_{0.7}Ca_{0.3}MnO_3$ (PCMO), it has been established that the growing Al metal electrode is oxidized whereas deposited PCMO thin film layer depicts the reduction process during deposition process as evidenced by thorough *in-situ* photoemission studies. Therefore, in interface type resistive switching process, the prominent electronic transport across interface is highly dependent upon the developed interfacial oxide thin layer. Such insulating oxide layers at interface are important in deciding the resistivity behavior during switching of the fabricated devices [17]. In TiN/SiO_2/Fe stacked structure, Feng *et al.* reported RS behavior induced because of thin FeO_x transition layer at SiO_2/Fe interface formed during processing of plasma-enhanced tetraethyl orthosilicate. However, after incorporating Pt into Fe electrode ($TiN/SiO_2/Fe_{0.73}Pt_{0.27}$) reduces the concentration of Fe in thin FeO_x layer which eventually improved the data dispersion of switching parameters [18]. In another report, the annealing of $TiN/SiO_2/FeO_x/$ FePt stratified structure dramatically modifies RS properties. Under optimized annealing conditions, excellent improvements have been observed including distinct reduction in RS parameters such asforming voltage, set/reset voltages, and their dispersions along with higher resistance i.e. ON/OFF ratio [19, 20].

3.2 Electrically induced redox reactions at the Interface

The conducting oxide e.g. manganites or cobaltites based memristive devices that employ reactive metal electrodes practically show uniform electrical conduction throughout the active area of device [9, 21]. It is known that the electrical resistance in case of the fabricated devices is low and increases primarily as a function of the applied bias. Such positive bias is applied onto the top electrode of the fabricated device. A careful analysis employing the cross-sectional TEM along with EELS and HAXPES measurements revealed that the actual thickness of oxide layer formed at or near the interface enhances noticeably due to the electroforming treatment [14, 17]. In $TiN/SiO_2/FeO_x/Fe$ device, Chang *et al.* demonstrated multilevel RS characteristics containing thin FeO_x transition layer which essentially assist in achieving controlled current compliance in SET process and stopped voltage during RESET process. Interestingly, the controlled external electric conditions facilitate tunable resistive states. The distinct mechanism for multilevel RS behavior has been realized by distinguishing the electrical behaviors, statistically which indicated the mobile-ion-assisted electrochemical redox governed RESET process [22].

3.3 Schottky-like metal/conducting oxide interfaces

Prior to explaining the resistive switching effects, the realization about the fundamental of the electronic properties related to conducting oxides is very important. Also, the current and voltage (*I-V*) behavior of interface formed at conducting oxide and metal electrode plays a key role in device performance. However, the contact resistance of such interfaces are largely modified by two major effects. One of the two effects is the existence of unwanted chemical reaction of metal electrode with conducting oxide. Secondly, the distinct Fermi level of conducting oxide and metal electrode leads to generation of space charge layer. The high contact resistance of interface is predominantly due to the Schottky barrier. This forms space charge region where essentially the majority charge carriers are depleted. In many cases of memory devices, the different electrode metals not only induce considerable modification in the resistive switching properties but also the contact resistance is affected significantly [6, 15, 23]. In case of stratified $M/Pr_{0.7}Ca_{0.3}MnO_3/$ $SrRuO_3$ (M/PCMO/SRO) and $M/SrTi_{0.99}Nb_{0.01}O_3/Ag$ (M/Nb:STO/Ag) where M is top electrode metal, the change in *I-V* characteristics has been discussed utilizing different electrode materials. The authors have used Ti and Au with work functions as ~4.3 and 5.1 eV, respectively whereas SRO possesses the highest value of work functions as 5.3 eV [15]. In present case, PCMO and Nb:STO exhibit only Ohmic contacts with SRO and Ag acting as the bottom metal electrode. It is known that while PCMO semiconducting oxide is dominated by *p*-type behavior, Nb:STO depicts *n*-type conduction. One can realize that the contact resistance between M and *p*-type dominated PCMO oxide is the largest for M with the least work function. Therefore, for PCMO based memory cells, Ti having the lowest work function demonstrates rectifying I-V behavior i.e. hysteretic characteristics distinctive to resistive switching properties. However, the contact resistance between M and *n*-type Nb:STO increases as the work function of M enhances. This indicates that Nb:STO based memory cells utilizing Au as the top electrode display hysteretic characteristics during *I-V* measurements. These observations are in good agreement with the fact that the rectifying behavior of *I-V* characteristics is governed by Schottky type barrier height formed at the interfaces. Therefore, an important and critical role of Schottky type barrier can be easily perceive in driving the resistive swathing effect in the fabricated memory cells. Considering the highly reactive nature of Ti top electrode material, it is also necessary to take the accounts of different chemical reactions occurring at the interface. For example, Ti being a more reactive metal can chemically react to the semiconducting oxide through the extraction of oxygen ions during film deposition and subsequent annealing procedure. Such events take place when the oxygen vacancies are injected *via* areas in close proximity to the interfaces. These oxygen vacancies are primarily of donor type and hence, capable of modifying the initial donor concentration of *n*-type Nb:STO semiconducting oxide. The considerable enhancement in the amount of donors ensures improved Ohmic contact conductance behavior. On the other hand, *p*-type PCMO oxide experiences the diminished conductivity which eventually results into an insulating type region of PCMO close to the interface. Such a phenomenon is similar to space charge effects which adds up to oxidize Ti metal. Thus, it evokes resistance at interface since the oxidation of Ti forming non-stoichiometric TiO_x has low conductivity.

I-V as well as *C-V* behavior demonstrate a hysteretic type characteristics if the resistance switching is area scalable. This is well explained on the basis of Schottky depletion model through the mechanism of electronic trapping or detrapping [24]. In practice, different contributions from the metal electrode work function, electron affinity of the *n*-type Nb:STO semiconducting oxide and interface trap sates residing within low-*k* interfacial layer are taken into consideration while estimating the accurate Schottky barrier height [25]. Interestingly, a noticeable sign related to electronic trapping or detrapping of bandgap states can be perceived within the depletion region for memory cells where resistive switching characteristics are area scalable. Also, due to small read out currents, a shorter retention time (in the range of 10^2–10^3 s) is measured for such cells which endorses the retention time and current variations obtained for only electronic switching. A voltage-induced unidirectional threshold resistive switching has been reported for Au/NiO/Nb:$SrTiO_3$ devices fabricated by pulsed laser deposition. Interestingly, only positive voltage values demonstrate the forming process controlled threshold resistive switching behavior [26].

Further, Schottky barrier at the interface is altered not only *via* the impact ionization but also depend onto the movement of oxygen vacancies driven applied electric field [15]. In another report, for single crystals of self-doped $SrTiO_3$ (STO), the authors have discussed the effect of electrode engineering with variable work function, device geometry and measurement configurations upon the resistive switching. Additionally, the various metal electrode combinations such as Ti and Pt has been exploited to analyze and manipulate the electrical transport i.e. Ohmic or Schottky type across junctions. It has been concluded that the observed resistive switching behavior is greatly influenced by changing the amount of oxygen vacancies only at or near the interface under an effective applied electrical bias [27]. For most of the switching binary or complex transition metal oxides (i.e. solid electrolytes), it is well known that current transport is the collective manifestation of electronic charge carriers as well as the mobile ions and related ionic defects. There exists cationic interstitials, delocalized electrons and oxygen vacancies in hypo-stoichiometric oxides whereas oxygen interstitials, delocalized holes and cationic vacancies are the major charge carriers in hyper-stoichiometric oxides.

In general, both the electron affinity of metal oxide materials and metal electrode work function define the Schottky barrier height. This implies the fact that the metal work function is believed to be directly related to the Schottky barrier. Nevertheless, in practice, the Schottky barrier height is considerably affected by the formation of metal electrode/oxide interfacial layer providing an additional capacitor i.e. insulator type thin slab. In this context, Cowley and Sze revealed that formation of such capacitive layers evoke a definite and noteworthy drop in voltage which ultimately alters the ideal height of Schottky barrier.

Moreover, the formation of interface between metal electrode and oxide layer is also described in terms of the Helmholtz plane and diffuse double layers [28]. It has been thoroughly discussed that there exists an intrinsic electrochemical potential difference at the interface of metal electrode and oxide which induces the transfer of electrons among oxide layer and metal. Such an event produces dipole layer at the interface due to the movement of electrons ensuing space charge effect. Under applied electrical bias, electronic charge carriers, ions and related ionic defects take part in screening the electric field via the diffuse double layer in the oxide region. However, there is a little screening of electric field at the metal electrode side because of sufficiently high enough concentration of electrons. It is evident that the screening length is extended much deeper inside the oxide layer than that of metal electrode owing to large difference in the concentration of charge carriers and electrons. While the first screening of electric field is caused by the ions residing over the Helmholtz plane, the second screening is primarily due to the electronic charge carriers, ions and related ionic defects present in the diffuse double layer [28].

For the most frequently employed metal oxides such as Ta_2O_5, HfO_2, and SiO_2 based memristive devices with compatible metal electrodes like Ta, Hf, and Ti have been extensively studied. For example, HfO_x/AlO_y-based homeothermic devices depict low-power and homogeneous RS behavior useful synaptic applications [29]. In vanadium-based devices, Lin *et al.* demonstrate excellent RS characteristics through interface where localized transition occurs [30]. Hsieh *et al.* discussed mitigation of critical issue of short-term relaxation in HfO_x/CeO_x derived RRAM devices [31]. In case of bilayer structure devices employing $Ni/SiN_x/HfO_2/p^{++}$-Si stacks, a self-rectifying has been shown to improve the sneak-path current emerging in the crossbar arrays fabrication process. Such bilayered devices provided enhanced rectification ratio usually $>10^4$. It has been revealed that during negative bias, the formation of large Schottky barrier of HfO_2 facilitate the reduction in current [32]. The progressive and consistent investigations have shown that interfacial layer exists inherently between the metal electrode and oxide. However, the oxygen affinity of metal as well as chemical and thermodynamic strength of the oxide layer determines the degree of oxidation of metal electrodes. Thus, the performance of fabricated device is highly dependent onto the extent of interfacial layer altering the initial crystal structure of oxide layer [33].

4. Tunnel barrier driven resistive switching

The tunnel ReRAM based on oxides is peculiarly known as the nonvolatile memory that demonstrate area dependent resistive switching properties. Unity Semiconductor Company first introduced the tunnel ReRAM and then Rambus Labs followed the work. ReRAM employs thin insulator type oxide layer to control the current through quantum-mechanical tunneling as magnetic RAM. In contrast to other oxide based RRAM system showing movement of oxygen ions, the quantum-mechanical tunneling effect within the insulator or barrier provides a more regulated current values in different processes such as SET, RESET and

Read-out which facilitate improved device performance. The mobile oxygen ion traveling across the insulator or barrier tunnel shows considerable impact onto the tunnel properties. Also, an appropriate change in thickness of tunnel barrier or oxide effectively helps into regulate the device current levels. Moreover, such a control over *I-V* characteristics for SET and RESET states offers transistor or diode-like selector less fabrication of passive cross-point configuration (1R) based working memory devices. This type of operation memory devices exhibit self-select feature. A precise and optimized tunnel device geometry with regulated current conduction serves high device yeild due to small variability in cell-to-cell and wafer-to-wafer operation. Further, the multibit storage can also be realized even in a single memory cell owing to an analog transition of the SET and RESET processes. The area dependent switching properties also enables smaller technology nodes in near future. Since tunnel ReRAM is still in development stage, it suffers from a few critical issues like ease of processing, integration and optimum retention time [34].

The tunnel ReRAM is also a stratified structure containing a very thin tunnel oxide layer (in the range of 2–3 nm only) placed between conducting oxide and metal electrode. The conducting meal oxide is usually a metallic type perovskite oxides including (Pr, Ca)MnO_3 or highly conducting La or Nb doped $SrTiO_3$. The stable electrical insulators possessing high-*k* oxides such as ZrO_2 or HfO_2 are employed generally as the tunnel oxide barrier layer [35]. The noble metals like Pt are utilized as electrodes for electrical biasing. The use of noble electrode metal is of utmost since they do not

Switching Device Fabrication Technique of Switching layer	Retention (sec)	Endurance (cycle)	Switching Voltage	Type of Switching	Ref
Ni/Si_3N_4/SiO_2/p + -Si low-pressure chemical vapor deposition (LPCVD)	—	—	V_{SET} = 4 V, V_{RESET} = −2 V,	Si_3N_4 (5 nm) as a resistive switching layer and SiO_2 (2.5 nm) tunnel Barrier	[36]
Pt/Ti/HfO_2/TiOx/Pt Atomic layer deposition (ALD)	10^4	$3x10^2$	—	HfO_2 (4 nm) as a resistive switching layer and TiO_x (6 nm) tunnel barrier	[37]
Pt/Ta_2O_5/TaO_x/TiO_2/Pt Atomic layer deposition (ALD)	10^{10}	10^4	—	TiO_2 (4 nm) as a resistive switching layer and Ta_2O_5(6 nm) tunnel barrier	[38]
Ti/HfO_2/Al_2O_3/TiN Atomic layer deposition (ALD)	—	—	—	HfO_2 (10 nm) as a resistive switching layer and Al_2O_3 (1 nm) tunnel barrier	[39]

Table 2.
Comparison for of switching parameters for tunnel barrier type RS devices.

react chemically with oxygen ions inducing formation of unwanted interfacial layer. The mechanism of tunnel ReRAM is radically different to interface based switching devices discussed above. In order to obtain good conductivity, both the conducing metal oxides e.g. (Pr, Ca)MnO_3 and tunnel oxide layer are processed at high temperature to ensure optimum crystallinity of the deposited films (**Table 2**) [34].

5. Ferroelectric resistive switching

In the late year *1970*, the first experimental work by Esaki reported the resistive switching behavior utilizing reversal of polarization in ferroelectric material [40]. This report followed many other ferroelectric based resistive switching materials such as the perovskitetitanates including $PbTiO_3$ [41], $BaTiO_3$ [42], and multiferroic system e.g. $BiFeO_3$ [43]. With consistent investigations, researchers have explained the resistive switching in ferroelectrics [40–43]. Usually, there exists two different types of ferroelectric based resistive switching memory. Considering the conduction mechanism, first is the ferroelectric tunneling junction and other is known as ferroelectric diode type memory. An ultrathin tunneling ferroelectric barrier is the primary component of the ferroelectric tunneling junctions. For such junctions, one can observe significant modification in tunnel oxide barrier height because of polarization reversal occurring in the ferroelectric materials. The ferroelectric tunnel barrier exhibits potential distribution of asymmetric type which evokes change in the height of barrier when reversal of polarization takes place. For memory cells, when both the metal electrodes are different that impose distinct screening lengths, the asymmetric potential distribution is achieved [44, 45]. The asymmetric behavior of potential has also been examined after placing a non-ferroelectric ultrathin layer onto the ferroelectric oxide layer [46]. The presence of dielectric ultrathin layer acts as a separation wall which essentially divides polarization charge in ferroelectric oxide layer and screening charge in metal electrode. Therefore, the non-ferroelectric i.e. dielectric layer possesses a certain distribution of the potential. In case of ferroelectric diode, the interface between metal electrode and ferroelectric layer observes the formation of Schottky type barrier. Upon the reversal of polarization in ferroelectric layer, the height of Schottky type barrier correspondingly changes (**Table 3**) [41].

Switching Device Fabrication Technique of Switching layer	Retention (sec)	Endurance (cycle)	Switching Voltage	Ref
Cu/Pb($Zr_{0.2}Ti_{0.8}$)O_3/ $La_{0.7}Sr_{0.3}MnO_3$/$SrTiO_3$	—	—	V_{SET} = 0.8 V, V_{RESET} = −2.1 V,	[45]
2-nm C-AFM Tip/BTO/30-nm LSMO ($La_{2/3}Sr_{1/3}MnO_3$)	—	—	V_{SET} = 3.5 V, V_{RESET} = −3.5 V	[42]
Ag/$BiFeO_3$/Ag	—	—	V_{Pulse} = 150 V, −150 V	[43]

Table 3.
Comparison of switching voltages for ferroelectric type RS devices.

6. Complementary switching

The complementary resistive switching (CRS) is the switching mode in which two distinct locations of switching are found. CRS devices contain variable concentration of defects within the switching layer. In particular, SET process occurs if the oxygen vacancies accumulate at either of metal electrodes. This takes place on account of the reduced oxygen vacancy concentration at the other end [47]. It states that both the interfaces cannot switch simultaneously. One needs to switch OFF when the other is ON. The mechanism of CRS is explained in terms of the distribution of defect within the insulating i.e. switching layer. After the applied electrical bias (positive or negative), the movement of defects towards certain electrode is observed. Depending upon the accumulation of defects at one or opposite interface, the SET and RESET states occur giving current levels under +ve and/or -ve bias conditions. With the help of either current compliance condition or sudden stop of bias at current peaks, only the SET state is obtained and one can limit the RESET state to takes place giving rise to bipolar switching [48]. Compared to other resistive switching devices, CRS provides slow and steady SET operation at one interface rather than an abrupt transition in current level. At the same time, RESET operation occurring at other interface eliminates the need of positive feedback [49]. It is known that an individual VCM device can exhibit CRS operation. For VCM devices with symmetric stack and similar top and bottom metal electrodes shows CRS phenomenon (**Table 4**).

Switching Device	Retention (sec)	Endurance (cycle)	Switching Voltage	Reason for Switching	Ref
$HfO_2/Al_2O_3/TiO_x$(HAT)	10^4	10^3	$V_{SET} = -3.8$ V, $V_{RESET} = -3.5$ V,	Anionic redistribution in HfO_2 and TiO_x layers, leaving Al_2O_3 as tunnel barrier	[50]
TiN/HfOx/Pt	10^4	10^3	$V_{SET} = -0.86$ V, $V_{RESET} = 1.08$ V,	BRS of device transforms to CRS after transitional processes through controlling compliance current	[51]
Pd/Ta_2O_{5-x}/TaO_y/Pd	~3.5x10^3	2x10^3	$V_{SET} = -0.9$ V, $V_{RESET} = 1.1$ V	Different oxygen compositions of Ta_2O_{5-x} and TaO_y layers, oxygen vacancies (V_O) can be exchanged between two layers.	[52]
Ta/ZnSnO/TiN	10^4	2x10^3	$V_{SET} = -2$ V, $V_{RESET} = 1.5$ V	TaO and TiON interface layers are formed at the top Ta/ZnSnO and bottom ZnSnO/TiN interfaces	[53]

Table 4.
Comparison for complementary type RS devices.

7. Interface controlled resistance (ICR)

The redox reaction driven resistive switching is dominated by the mechanism in which the charged defects or impurities rearrange at the metal electrode and oxide interface. Such a rearrangement of mobile defects reduce the contact resistance due to large interface charge concentration which ultimately decreases the contact barrier height as well as width. This process is examined through the Mott–Schottky theory dealing with the contact formed among metal and non-metal elements. For nanoionic devices, the resistance switching mechanism based on the formation of an interfacial layer has been investigated in a wide variety of metal oxides including TiO_2 [54], HfO_2 [55, 56], ZrO_2 [57, 58] Ta_2O_5 etc. It is well known that the metal oxides are more prone to oxygen ion i.e. ease of defect formation and hence, largely show nonstoichiometric nature. Therefore, the inherent nonstoichiometric nature of metal oxides produce several lattice point defects or impurities like vacancies and/or interstitials of metal and oxygen ions. Such defects present in metal oxides can be similar to acceptors or donors and are capable of modifying the electronic properties and the switching characteristics of device. Owing to sufficient concentration of charge carriers i.e. defects, such metal oxides can act equivalent to extrinsic semiconductor and thus, the classical semiconductor model governs the electronic conduction.

In order to highlight three main differences between nonstoichiometric metal oxide and classical semiconductor, researchers categorized them into a particular group known as mixed ionic–electronic conductors or chemiconductors [59, 60]. Followings are the distinctions used for classification:

i. the variation in oxide composition i.e. nonstoichiometric feature induces defect ions producing donors and acceptors i.e. *dopants* in chemiconductors,

ii. under applied electrical bias, while the defect ions are able mobile within the lattice, dopants in semiconductors are fixed at certain positions in lattice, and.

iii. however, the dopants are non-uniformly distributed at the interface, in particular.

Apart from the above differences, the classical semiconductor acts similar to chemiconductors possessing sufficient nonstoichiometric feature. Therefore, their electrical characteristics are described in the framework of the Mott–Schottky model producing Schottky barriers at the interface (**Table 5**).

Switching Device	Retention (sec)	Endurance (cycle)	Switching Voltage	Reason for Switching	Ref
Al/MoO_x/Pt	—	5x10^2	V_{SET} = −3 V, V_{RESET} = −2 V,	Self-rectifying and interface-controlled	[61]
TiN/ZrO_2/Pt	120	—	—	oxygen vacancy conducting filamentary paths	[57]
s-In/ NSTO/o-In	10^4	3.5x10^3	—	Schottky interface	[62]

Table 5.
Comparison for interface controlled resistance RS devices.

8. Effect of grain surface area and grain boundary

Interfacial type resistive switching (RS) in memristive device is dominated by the surface morphology and properties. Under the application of an external voltage bias, the density distribution of defects/vacancies can vary along the film thickness, which affects the bandgap of material and produces unpredictable behavior in resistive switching response. These uncertain changes at the interface, barrier bandgap leads to a change in the sample's resistance in an interfacial RS which is extremely sensitive to interfacial properties.

Generally, interface is defined between two different materials systems but sometimes interface can also be defined at the regions which have the same composition and crystal structure but different crystal orientations (even inside the same solid such as grain boundaries (GBs)-including tilt and twist boundaries-twin boundaries and stacking faults) [34]. Moreover, an interface is affected by the several external environment conditions such as air, vacuum, moisture and some other material properties such as the crystallinity of solids at the interface. However, the defect chemistry is comparatively different in the proximity of a charged interface (GB) from the bulk situation (single crystal). Basically, charged interface induces the redistribution of the mobile charge carriers in the space-charge layer region while in bulk, the electroneutrality has been played an important role (at equilibrium) between differently charged point defects. In this section, the effect of grain boundaries (GBs) and grain surfaces (GSs) on RS have discussed.

Oxygen vacancies and defects are considered responsible for resistive switching phenomena in oxides materials. For nanoscaled materials, GBs conductivity is directly proportional to grain size, and it may modulate according to the direction of current flow (perpendicular and parallel direction of GBs) [63]. However, the position of the GBs is dependent on the shape and size of the grain J. Maier [63] has reported that in case of yttria-doped zirconia if the grain size decreases at particular dimension (~ 50 nm in diameter) then most of the current passes perpendicular to the GB axis and the conductivity parallel to GB becomes negligible. Further, the formation of GBs when two adjacent and equally oriented grains are rotated to each other and twisted GB occurs when the rotation axis is perpendicular to the boundary. On the other hand, if the rotation axis is lied in the boundary plane, a tilted boundary is resulted [34]. Moreover, the degree of rotation also affects the coherence of the final grain boundary. GB with minimum rotation angle can be treated as an group of edge dislocations and aggregation of screw dislocations [34]. The concentration of oxygen vacancies (O_v) is comparatively higher at GBs which leads to leakage current flow through the GBs predominantly [64]. Any variation in O_v will be closely related to grain boundary and grain surface area. In some polycrystalline oxide thin film, structural defects, grain boundaries and local nonstoichiometric regions are responsible for high leakage current. Further, high electrical stress due to applied electrical potential, induces traps/cracks along the GBs. Induced traps/cracks also increases the leakage current and size of the conduction region at the GBs as compared to the grain regions [65]. In recent years, the ab initio calculations and conductive atomic force microscopy (CAFM) have demonstrated to study the charge transport through grain boundaries in polycrystalline HfO_2 [66].

The space-charge conduction model for acceptor-doped zirconia suggests that the lower ionic conductivity in zirconia occurs due to the depletion of oxygen vacancies and excess the positive charge laying in the GB core [67, 68]. However, dislocations appeared in YSZ single crystal due to plastic deformation does not improve the material's electrical transport significantly [69]. Further, the resistive switching in WO_3 thin film is dominated by the grain surface region, not by the GB [70].

In case of polycrystalline oxide films, GBs contain a high density of defects, which will accumulate the more traps inside the grain regions. A traps present inside the grain may cause a percolation path under a high electric field [65]. In filamentary type conduction, oxygen vacancies/ions would form a filaments throughout the oxide layer (highly conductive path) via GB [71]. Another interpretation of conductive filament's formation is connected with the motion of the O^{2-} ions which usually appears near the crystal defects such as oxygen vacancies and GBs [56, 72, 73]. A filament forming behavior in switching oxide film can be controlled by controlling the grain size underneath the top electrode and smaller grain size indicate the large number of GBs. These types of correlation can be identified by varying the electrode size and the number of GB underneath. However, these correlation becomes are not impactful if electrode size less than the individual grain size [74]. In addition, Das *et al*. have discussed the effect of GBs, GSs, and surface morphology such as (hillocks, lattice mismatch) on the statistical variation of RS parameters (forming voltage, set-reset voltage) in yttria-based resistive switching device (**Figure 2**) [64]. Successive RS operations depend on the inhomogeneous changes in defect structure, and as a result, the switching parameters also vary persistently. During RS process, there are several type of sources in different oxides which provoke variability in device parameters. However, the formation and recombination of oxygen vacancies is highly stochastic in nature and play dominant role in deciding degree of variability. After analyzing the experimental data, Monte Carlo simulation has established a potential stochastic model that relates subsequent RS behavior to the initial states of contact in resistive memory cells [75].

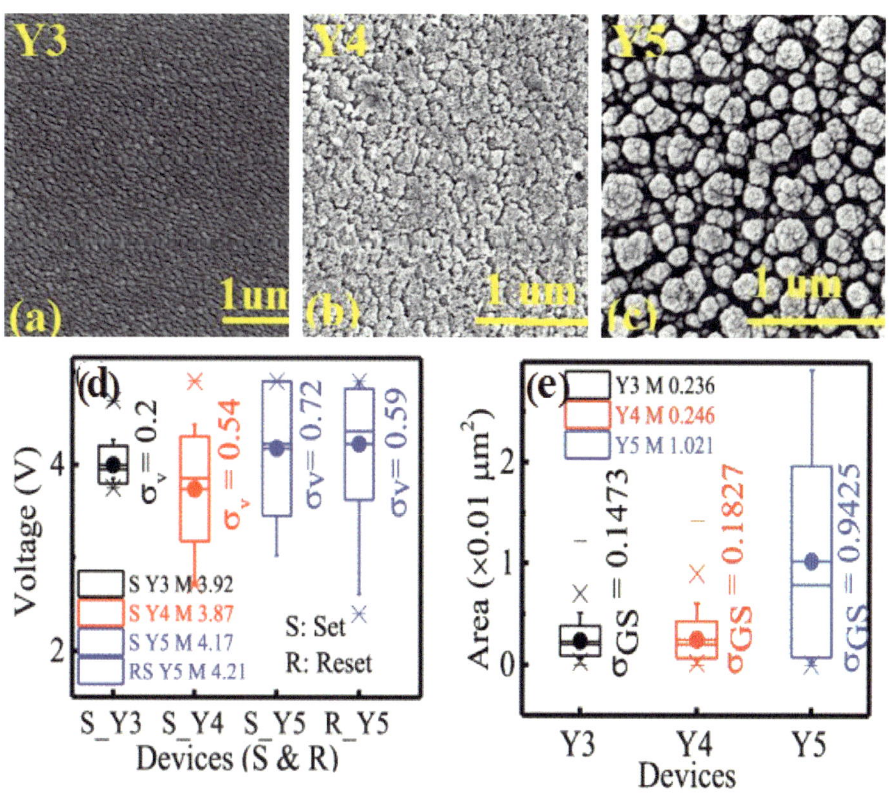

Figure 2.
Yttria layer (~80 nm) is deposited at different substrate temperatures of 300 (Y3), 400 (Y4), and 500°C (Y5) by dual ion beam sputtering system. Scanning electron microscope (FESEM) images of (a) Y3 (b) Y4 (c) Y5. (d) Mean (M) and standard deviation (σ) of the set and reset voltages. (e) Mean (M) and standard deviation (σ) of the grain surface area (figure d and e. reprinted with permission: Ref. [64]).

9. Conclusions

This chapter dealt with different types of resistive switching where the role of interface formation modifying the switching properties was thoroughly discussed. The surface properties of thin films demonstrated significant dependence onto the predominant factors e.g. redox reactions, interface formation along with tunnel barrier thereby affected the device performance. The morphological characteristics of surface across the interface containing GBs and GSs regulated the charge carrier transport due to modified band structure. Owing to relatively high defect concentration at the interface, GBs controlled the leakage current behavior of device.

Conflict of interest

This book chapter is written through contributions of all authors. All authors have given approval to the final version of the manuscript. Mangal Das and Sandeep Kumar is the main author and co-author, respectively. The authors declare no competing financial interest.

Author details

Mangal Das* and Sandeep Kumar
Department of Electrical Engineering, Indian Institute of Technology Bombay, Mumbai, India

*Address all correspondence to: mangaldas111@iitb.ac.in

References

[1] Baikalov, A., et al., *Field-driven hysteretic and reversible resistive switch at the Ag–Pr 0.7 Ca 0.3 MnO 3 interface.* Applied Physics Letters, 2003. **83**(5): p. 957-959.

[2] Chen, X., et al., *Direct resistance profile for an electrical pulse induced resistance change device.* Applied Physics Letters, 2005. **87**(23): p. 233506.

[3] Seong, D.-j., et al., *HPHA Effect on Reversible Resistive Switching of Pt/ Nb-Doped SrTiO3 Schottky Junction for Nonvolatile Memory Application.* Electrochemical and Solid State Letters, 2007. **10**(6): p. H168.

[4] Nian, Y., et al., *Evidence for an oxygen diffusion model for the electric pulse induced resistance change effect in transition-metal oxides.* Physical review letters, 2007. **98**(14): p. 146403.

[5] Yang, M.K., et al., *Bipolar resistive switching behavior in Ti/MnO 2/Pt structure for nonvolatile memory devices.* Applied Physics Letters, 2009. **95**(4): p. 042105.

[6] Sawa, A., et al., *Hysteretic current–voltage characteristics and resistance switching at a rectifying Ti/ Pr 0.7 Ca 0.3 Mn O 3 interface.* Applied Physics Letters, 2004. **85**(18): p. 4073-4075.

[7] Rozenberg, M., I. Inoue, and M. Sanchez, *Nonvolatile memory with multilevel switching: a basic model.* Physical review letters, 2004. **92**(17): p. 178302.

[8] Oka, T. and N. Nagaosa, *Interfaces of correlated electron systems: proposed mechanism for colossal electroresistance.* Physical review letters, 2005. **95**(26): p. 266403.

[9] Hasan, M., et al., *Uniform resistive switching with a thin reactive metal interface layer in metal-La 0.7 Ca 0.3 Mn O 3-metal heterostructures.* Applied Physics Letters, 2008. **92**(20): p. 202102.

[10] Sim, H., et al. *Excellent resistance switching characteristics of Pt/SrTiO/sub 3/schottky junction for multi-bit nonvolatile memory application*. in *IEEE InternationalElectron Devices Meeting, 2005. IEDM Technical Digest.* 2005. IEEE.

[11] Meunier, B., et al., *Microscopic Mechanisms of Local Interfacial Resistive Switching in LaMnO3+ δ.* ACS Applied Electronic Materials, 2019. **1**(5): p. 675-683.

[12] Lähteenlahti, V., et al., *Transport properties of resistive switching in Ag/Pr0. 6Ca0. 4MnO3/Al thin film structures.* Journal of Alloys and Compounds, 2019. **786**: p. 84-90.

[13] Lv, F., et al., *Multilevel resistance switching behavior in PbTiO3/Nb: SrTiO3 (100) heterostructure films grown by hydrothermal epitaxy.* Journal of Alloys and Compounds, 2019. **778**: p. 768-773.

[14] Asanuma, S., et al., *Relationship between resistive switching characteristics and band diagrams of Ti/Pr 1- x Ca x MnO 3 junctions.* Physical Review B, 2009. **80**(23): p. 235113.

[15] Sawa, A., *Resistive switching in transition metal oxides.* Materials today, 2008. **11**(6): p. 28-36.

[16] Kawano, H., et al., *Enhancement of switching capability on bipolar resistance switching device with Ta/Pr0. 7Ca0. 3MnO3/Pt structure.* Applied physics express, 2008. **1**(10): p. 101901.

[17] Borgatti, F., et al., *Chemical insight into electroforming of resistive switching manganite heterostructures.* Nanoscale, 2013. **5**(9): p. 3954-3960.

[18] Feng, L.-W., et al., *A study of resistive switching effects on a thin FeO x transition*

layer produced at the oxide/iron interface of TiN/SiO 2/Fe-contented electrode structures. Applied Physics Letters, 2010. **96**(5): p. 052111.

[19] Feng, L.-W., et al., *Improvement of resistance switching characteristics in a thin FeO x transition layer of TiN/SiO 2/ FeO x/FePt structure by rapid annealing.* Applied Physics Letters, 2010. **96**(22): p. 222108.

[20] Chang, Y.-F., L.-W. Feng, and T.-C. Chang, *Mechanism and characterizations studies of resistive switching effects on a thin FeOx-transition layer of the Ti/TiN/ SiO2/FeOx/FePt structure by thermal annealing treatments.* Materials chemistry and Physics, 2011. **131**(1-2): p. 262-267.

[21] Meyer, R., et al. *Oxide dual-layer memory element for scalable non-volatile cross-point memory technology*. in *2008 9th Annual Non-Volatile Memory Technology Symposium (NVMTS)*. 2008. IEEE.

[22] Chang, Y.-F., T.-C. Chang, and C.-Y. Chang, *Investigation statistics of bipolar multilevel memristive mechanism and characterizations in a thin FeOx transition layer of TiN/SiO2/FeOx/Fe structure.* Journal of applied physics, 2011. **110**(5): p. 053703.

[23] Tokunaga, Y., et al., *Colossal electroresistance effect at metal electrode/La 1- x Sr 1+ x Mn O 4 interfaces.* Applied physics letters, 2006. **88**(22): p. 223507.

[24] Park, C., et al., *Electrode-dependent electrical properties of metal/Nb-doped Sr Ti O 3 junctions.* Journal of Applied Physics, 2008. **103**(5): p. 054106.

[25] Mikheev, E., et al., *Resistive switching and its suppression in Pt/Nb: SrTiO 3 junctions.* Nature communications, 2014. **5**(1): p. 1-9.

[26] Guo, M., et al., *Unidirectional threshold resistive switching in Au/NiO/ Nb: SrTiO3 devices.* Applied Physics Letters, 2017. **110**(23): p. 233504.

[27] Peng, H., et al., *Effects of electrode material and configuration on the characteristics of planar resistive switching devices.* APL Materials, 2013. **1**(5): p. 052106.

[28] Jeong, D.S., et al., *Emerging memories: resistive switching mechanisms and current status.* Reports on progress in physics, 2012. **75**(7): p. 076502.

[29] Kim, S., et al., *Neuronal dynamics in HfO x/AlO y-based homeothermic synaptic memristors with low-power and homogeneous resistive switching.* Nanoscale, 2019. **11**(1): p. 237-245.

[30] Lin, C.-Y., et al., *A comprehensive study of enhanced characteristics with localized transition in interface-type vanadium-based devices.* Materials Today Physics, 2020. **13**: p. 100201.

[31] Hsieh, C.-C., et al., *Short-Term Relaxation in HfO x/CeO x Resistive Random Access Memory With Selector.* IEEE Electron Device Letters, 2017. **38**(7): p. 871-874.

[32] Cho, H., et al., *Bipolar resistive switching with unidirectional selector function in nitride/oxide heterostructures.* Journal of Physics D: Applied Physics, 2020. **53**(43): p. 435102.

[33] Cho, D.-Y., et al., *Interfacial metal–oxide interactions in resistive switching memories.* ACS applied materials & interfaces, 2017. **9**(22): p. 19287-19295.

[34] Ielmini, D. and R. Waser, *Resistive switching: from fundamentals of nanoionic redox processes to memristive device applications*. 2015: John Wiley & Sons.

[35] Kumar, S., S. Rai, and C. Rath, *Multifunctional role of dysprosium in HfO 2: stabilization of the high temperature cubic phase, and magnetic and photoluminescence properties.* Physical Chemistry Chemical Physics, 2017. **19**(29): p. 18957-18967.

[36] Kim, S., et al., *Resistive switching characteristics of Si3N4-based resistive-switching random-access memory cell with tunnel barrier for high density integration and low-power applications.* Applied Physics Letters, 2015. **106**(21): p. 212106.

[37] Lee, S., et al. *Effect of TiO x-based tunnel barrier on non-linearity and switching reliability of resistive random access memory*. in *2014 14th Annual Non-Volatile Memory Technology Symposium (NVMTS)*. 2014. IEEE.

[38] Woo, J., et al. *Multi-layer tunnel barrier (Ta 2 O 5/TaO x/TiO 2) engineering for bipolar RRAM selector applications*. in *2013 Symposium on VLSI Technology*. 2013. IEEE.

[39] Chand, U., et al., *Mechanism of nonlinear switching in HfO 2-based crossbar RRAM with inserting large bandgap tunneling barrier layer.* IEEE Transactions on Electron Devices, 2015. **62**(11): p. 3665-3670.

[40] Esaki, a.L., R. Laibowitz, and P. Stiles, *Polar switch.* IBM Tech. Discl. Bull, 1971. **13**(2161): p. 114.

[41] Blom, P., et al., *Ferroelectric schottky diode.* Physical Review Letters, 1994. **73**(15): p. 2107.

[42] Garcia, V., et al., *Giant tunnel electroresistance for non-destructive readout of ferroelectric states.* Nature, 2009. **460**(7251): p. 81-84.

[43] Choi, T., et al., *Switchable ferroelectric diode and photovoltaic effect in BiFeO3.* Science, 2009. **324**(5923): p. 63-66.

[44] Zhuravlev, M.Y., et al., *Giant electroresistance in ferroelectric tunnel junctions.* Physical Review Letters, 2005. **94**(24): p. 246802.

[45] Pantel, D., et al., *Room-temperature ferroelectric resistive switching in ultrathin Pb (Zr0. 2Ti0. 8) O3 films.* ACS nano, 2011. **5**(7): p. 6032-6038.

[46] Zhuravlev, M.Y., et al., *Tunneling electroresistance in ferroelectric tunnel junctions with a composite barrier.* Applied Physics Letters, 2009. **95**(5): p. 052902.

[47] Marchewka, A., R. Waser, and S. Menzel. *Physical modeling of the electroforming process in resistive-switching devices*. in *2017 International Conference on Simulation of Semiconductor Processes and Devices (SISPAD)*. 2017. IEEE.

[48] Nardi, F., et al., *Complementary switching in oxide-based bipolar resistive-switching random memory.* IEEE transactions on electron devices, 2012. **60**(1): p. 70-77.

[49] Brivio, S., J. Frascaroli, and S. Spiga, *Role of metal-oxide interfaces in the multiple resistance switching regimes of Pt/HfO2/TiN devices.* Applied Physics Letters, 2015. **107**(2): p. 023504.

[50] Banerjee, W., et al., *Design of CMOS compatible, high-speed, highly-stable complementary switching with multilevel operation in 3D vertically stacked novel HfO2/Al2O3/TiOx (HAT) RRAM.* Advanced Electronic Materials, 2018. **4**(2): p. 1700561.

[51] Chen, X., et al., *Complementary resistive switching behaviors evolved from bipolar TiN/HfO2/Pt device.* Applied Physics Letters, 2016. **108**(5): p. 053504.

[52] Yang, Y., P. Sheridan, and W. Lu, *Complementary resistive switching in tantalum oxide-based resistive memory devices.* Applied Physics Letters, 2012. **100**(20): p. 203112.

[53] Ismail, M., et al., *Bipolar, complementary resistive switching and synaptic properties of sputtering deposited ZnSnO-based devices for electronic synapses.* Journal of Alloys and Compounds, 2021. **862**: p. 158416.

[54] Yang, J.J., et al., *Memristive switching mechanism for metal/oxide/metal nanodevices.* Nature nanotechnology, 2008. **3**(7): p. 429-433.

[55] Gonon, P., et al., *Resistance switching in HfO 2 metal-insulator-metal devices.* Journal of Applied Physics, 2010. **107**(7): p. 074507.

[56] Kumar, S. and C. Rath, *Oxygen Vacancy Mediated Stabilization of Cubic Phase at Room Temperature and Resistive Switching Effect in Sm-and Dy-Doped HfO2 Thin Film.* physica status solidi (a), 2020. **217**(1): p. 1900756.

[57] Sun, B., et al., *The effect of current compliance on the resistive switching behaviors in TiN/ZrO2/Pt memory device.* Japanese Journal of Applied Physics, 2009. **48**(4S): p. 04C061.

[58] Nishi, Y., et al., *Origin of the SET kinetics of the resistive switching in tantalum oxide thin films.* IEEE electron device letters, 2013. **35**(2): p. 259-261.

[59] Schultze, J.W. and M. Lohrengel, *Stability, reactivity and breakdown of passive films. Problems of recent and future research.* Electrochimica Acta, 2000. **45**(15-16): p. 2499-2513.

[60] Stimming, U. and J. Schultze, *A semiconductor model of the passive layer on iron electrodes and its application to electrochemical reactions.* Electrochimica Acta, 1979. **24**(8): p. 859-869.

[61] Hsu, C.-C., et al., *Self-rectifying and interface-controlled resistive switching characteristics of molybdenum oxide.* Journal of Alloys and Compounds, 2019. **779**: p. 609-617.

[62] Sun, J., et al., *Control of normal and abnormal bipolar resistive switching by interface junction on In/Nb: SrTiO3 interface.* Applied Physics Letters, 2012. **101**(13): p. 133506.

[63] Maier, J., *Ionic conduction in space charge regions.* Progress in solid state chemistry, 1995. **23**(3): p. 171-263.

[64] Das, M., et al., *Effect of surface variations on the performance of yttria based memristive system.* IEEE Electron Device Letters, 2018. **39**(12): p. 1852-1855.

[65] Shubhakar, K., et al., *Study of preferential localized degradation and breakdown of HfO2/SiOx dielectric stacks at grain boundary sites of polycrystalline HfO2 dielectrics.* Microelectronic engineering, 2013. **109**: p. 364-369.

[66] Pirrotta, O., et al., *Leakage current through the poly-crystalline HfO2: Trap densities at grains and grain boundaries.* Journal of Applied Physics, 2013. **114**(13): p. 134503.

[67] Guo, X. and Y. Ding, *Grain boundary space charge effect in zirconia: Experimental evidence.* Journal of the Electrochemical Society, 2003. **151**(1): p. J1.

[68] Guo, X., et al., *Role of space charge in the grain boundary blocking effect in doped zirconia.* Solid State Ionics, 2002. **154**: p. 555-561.

[69] Otsuka, K., et al., *Effects of dislocations on the oxygen ionic conduction in yttria stabilized zirconia.* Materials transactions, 2004. **45**(7): p. 2042-2047.

[70] Lee, J.S., S. Lee, and T.W. Noh, *Resistive switching phenomena: A review of statistical physics approaches.* Applied Physics Reviews, 2015. **2**(3): p. 031303.

[71] Liu, X., et al., *Resistive switching mechanism of a Pr0. 7Ca0. 3MnO3-based memory device and assessment of its suitability for nano-scale applications.* Journal of the Korean Physical Society, 2011. **59**(2): p. 497-500.

[72] Di Ventra, M. and Y.V. Pershin, *Memory materials: a unifying description.* Materials Today, 2011. **14**(12): p. 584-591.

[73] Pan, F., et al., *Recent progress in resistive random access memories: materials, switching mechanisms, and performance.* Materials Science and Engineering: R: Reports, 2014. **83**: p. 1-59.

[74] Govoreanu, B., et al. *10× 10nm 2 Hf/ HfO x crossbar resistive RAM with excellent performance, reliability and low-energy operation*. in *2011 International Electron Devices Meeting*. 2011. IEEE.

[75] Kao, Y.-F., et al., *A study of the variability in contact resistive random access memory by stochastic vacancy model.* Nanoscale research letters, 2018. **13**(1): p. 1-10.

Chapter 3

Artificial Synapses Based on Atomic/Molecular Layer Deposited Bilayer-Structured Memristive Thin Films

Chang Liu, Lin Zhu, Lai-Guo Wang and Ai-Dong Li

Abstract

This chapter deals with several kinds of ultrathin bilayer-structured memristors, such as Pt/Al_2O_3/HfO_2/TiN, Pt/HfO_2/HfO_x/TiN, Pt/TiO_2/Ti-based maleic acid (Ti-MA)/TaN, among which the asymmetric memristive functional layers were designed and prepared by atomic layer deposition (ALD) or molecular layer deposition (MLD) technique. These bilayer memristors exhibit a typical bipolar resistive switching characteristic, in accord with the space charge limited current model. Some important biologic synaptic functions have been achieved, including nonlinear transmission characteristics, spike-timing-dependent plasticity, short−/long-term plasticity, paired-pulse facilitation, and conditioned reflex. The mechanism of bilayer memristive device has been proposed based on oxygen vacancies migration/diffusion model. Above all the ultrathin bilayer memristors fabricated by low temperature ALD/MLD are one competitive candidate for neuromorphic simulation and flexible electronic applications.

Keywords: memristor, atomic layer deposition, bilayer, synapse, mechanism

1. Introduction

The memristor concept was first proposed as the fourth fundamental passive circuit element by Chua in 1971 based on the completeness of the circuit theory, which indicates the relationship between magnetic flux and charge [1, 2]. After thirty seven years, Strukov et al. eventually found the missing memristor in studying TiO_2 cross-arrays in 2008 [2]. This draws the extensive and intensive attention from the academia and the industry. Memristor is a two-terminal electrical device whose resistance can be tuned by changing the flux or charge through it. Memristor possesses a lot of advantages, e.g., simple device architecture, high energy efficiency, better compatibility with semiconductor industry, and high integration density.

A neural synapse, as the basic unit of learning and memory in the brain, plays a critical role in biological neural networks. Electronic synapses are utilized to emulate the bio-synapses' functions. Some researches on synapse simulation have been reported by adjusting synaptic weights so as to make an effective bio-inspired computing system [3–6]. Nevertheless, most work chose transistors and capacitors to realize artificial synapse, which produced high energy consumption at high integration density and limited the programming running. The new memristor has nonlinear transfer characteristics similar to the bio-synapse and is regarded as the closest to the synaptic device [4].

Although various materials and structures exhibit memristive behavior, almost all the memristor systems are based on the structural asymmetry [7, 8]. For example, in the metal–insulator–metal (MIM) structure, the defects such as oxygen vacancy or active ions in the insulator layer can induce structural asymmetry under the action of the external field, or when one of the metal electrodes is active. Therefore, the asymmetric bilayer-structured memristors play a crucial role in constructing artificial neural networks for brain-inspired applications.

Atomic layer deposition (ALD) is a kind of commercial technology compatible with semiconductor processing. It shows unusual advantages in controllable fabrication of nano-laminate thin films due to its unique sequential self-limiting surface reaction mechanism at low growth temperature [9, 10]. In early 2001 ALD has been known as candidate technology preferred for semiconductor industry along with metalorganic chemical vapor deposition (MOCVD) and plasma-enhanced CVD by the international technology roadmap for semiconductors (ITRS) [11]. ALD has become one of the most competitive deposition techniques for microelectronics and nanotechnology owing to sub-nanometer thickness control, large-area uniformity, excellent three-dimensional conformality, and good reproducibility. Thin films with low defect density can be prepared by ALD even at room temperature (RT) with plasma assistance [12]. Evidently, low temperature or RT ALD technology can greatly widen the flexible substrate choice range, showing exciting potentials in flexible electronic device fabrication. Molecular layer deposition (MLD) can be regarded as the subtype of ALD due to the molecular nature of the deposition process, suitable for growth of organic–inorganic hybrid materials [13].

In this section, we fabricated several synaptic devices of asymmetric bilayer-structured ultrathin memristors by atomic layer deposition (ALD) and molecular layer deposition (MLD), such as Pt/AlO_x/HfO_x/TiN, Pt/HfO_2/HfO_x/TiN, Pt/TiO_2/Ti-based maleic acid (Ti-MA)/TaN. Some biological synapse-like functions of long−/short-term plasticity (LTP and STP), spike-timing-dependent plasticity (STDP), and paired-pulse facilitation (PPF) have been achieved simultaneously. A memristive mechanism of an asymmetric bilayer-structured synaptic device has been proposed to explain synaptic plasticity based on the oxygen vacancy migration/diffusion model.

2. Bilayer-structured ultrathin memristors

2.1 Fabrication processing

Asymmetric bilayer-structured ultrathin memristor based on Pt/A/B/TiN or TaN was fabricated on SiO_2/Si substrates by thermal-ALD (TALD), MLD and plasma-enhanced ALD (PEALD), as illustrated in **Figure 1a**. Herein A and B act as asymmetric memristive functional layer, PEALD TiN or sputtered TaN as bottom electrode, sputtered Pt as top electrode with a spot size in diameter of 150 μm. **Table 1** gives several typical bilayer ultrathin memristors and their architectures. The related deposition conditions have been listed in **Table 2**, including used metal precursors and reactants, source temperature and deposition temperature, and growth per cycle (GPC).

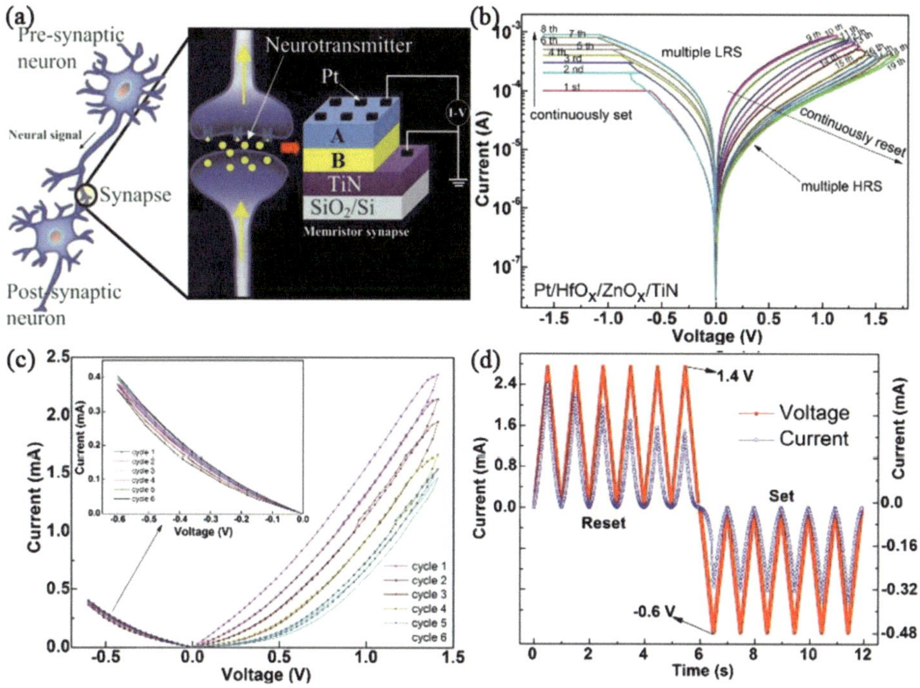

Figure 1.
(a) Diagram of the asymmetric bilayer-structured ultrathin memristor. (b) I–V characteristics of the Pt/HfO_x/ZnO_x/TiN synapse device measured by a modified DC double ramp sweep. The sweep sequence is denoted by the number. (c) I–V characteristics of the memristor at positive and negative bias voltages. The voltage sweep range is from 0 to 1.4 (−0.6) V then back to 0 V, and the time for a sweep cycle is 1 s. the device conductivity continuously decreases or increases during the positive or negative voltage sweeps. (d) the curves of voltage and current versus time, which are plotted from the data in (c) [15].

Device structure	A (thickness)	B (thickness)
Pt/HfO_x/ZnO_x/TiN	HfO_x (5 nm)	ZnO_x (5 nm)
Pt/AlO_x/HfO_x/TiN	AlO_x (5 nm)	HfO_x (5 nm)
Pt/TiO_2/Ti-MA/TaN	Ti-MA (4 nm)	TiO_2 (4 nm)

Table 1.
Several typical bilayer-structured ultrathin memristors and their architectures.

	Material	Metal precursor	Precursor temperature	Reactant	Deposition temperature	GPC (Å/cycle)
TALD	HfO_2	TEMAH	155°C	H_2O	250°C	1
	AlO_x	TMA	RT	H_2O	250°C	1
	ZnO_x	DEZ	RT	H_2O	250°C	1.3
	TiO_2	$TiCl_4$	RT	H_2O	250°C	0.3
PEALD	HfO_x	TEMAH	155°C	H_2 plasma	250°C	1
	TiN	$TiCl_4$	RT	NH_3 plasma	400°C	0.5
MLD	Ti-MA	$TiCl_4$	RT	MA(135°C)	160°C	1.4

Table 2.
Deposition conditions of asymmetric functional layers in memristors prepared by TALD/PEALD/MLD. Here TEMAH, TMA, DEZ, and MA refer to $Hf[N(C_2H_5)CH_3]_4$, $Al(CH_3)_3$, $Zn(C_2H_5)_2$, maleic acid, respectively.

2.2 Electrical performances and synaptic functions

The electrical properties were measured under DC sweep and pulse modes using semiconductor parameter analyzer on probe station. The bottom electrode of memristors was set on ground and all the voltage signals were applied to the top electrode. The asymmetric bilayer ultrathin memristors were exploited to mimic some important synaptic functions such as long-term potentiation/depression, the transition from STP to LTP, PPF and STDP.

2.2.1 TiN/ZnO$_x$/HfO$_x$/Pt inorganic memristor

The *I–V* curves of Pt/HfO$_x$/ZnO$_x$/TiN inorganic memristor are plotted in **Figure 1b** under a modified DC double sweep. To mimic the functions of a nerve synapse, one multiple-state resistances should be obtained in bilayer memristor. A continuous set or reset process was performed by successive increasing the compliance from 0.1 to 1.0 mA at an interval of 0.1 mA or altering the reset voltage from 1.0 to 1.7 V at an interval of 0.05 V. 8 low resistance states (LRS) and 11 distinguishable high resistance states (HRS) are observed during consecutive set and reset process, respectively. Moreover the resistance can be continuously reduced or raised between multiple intermediate states without going back to the original state, which is key for electronic synapse [14]. The device conductivity decreases continuously with six easily recognized states after exerting sweep positive bias voltage from 0 to 1.4 V six times and the elevated conductivity with difficultly distinguishable ones after sweep negative pulse voltage from 0 to −0.6 V (**Figure 1c** and **d**), indicating the conductance change caused by consecutive potentiating or depressing signals. It can be attributed to the dynamic change of oxygen vacancy concentration and distribution in asymmetric bilayer structure of HfO$_x$/ZnO$_x$ under various electrical signals [15].

A series of pulse signals were designed and applied to the memristor to test the important STDP rule in the Hebbian learning theory, as seen in the insets of I and III of **Figure 2**, including the V−/V+ = −1.0 V/1.0 V pulse pair signal as a presynaptic and postsynaptic spike with the 3 s interval time. Such design can prevent from the disturbance of excitatory postsynaptic current [16]. The time interval between the final presynaptic spike and the initial postsynaptic spike is defined as the relative time of Δt. The relative change of the synaptic weights (ΔW) is defined as:

$$\Delta W = (I_2 – I_1)/I_1 \times 100\% \tag{1}$$

The initial postsynaptic or presynaptic current I_1 was used as the control value. After the spike pair was applied and over for 5 min, the measured presynaptic or postsynaptic current was I_2.

The dependence of ΔW *on* Δt *of* Pt/HfO$_x$/ZnO$_x$/TiN in **Figure 2** II and IV follows the STDP learning rule. While the presynaptic spike happens before the postsynaptic spike ($\Delta t < 0$), synaptic weights enhance, indicating long-term potentiation (LTPo); while the presynaptic spike appears after the postsynaptic spike, synaptic weights become small ($\Delta t > 0$), implying long-term depression (LTD). And the shorter the Δt between the two spikes, the larger the ΔW. The STDP data points of memristor in **Figure 2** show evident statistical scatter, similar to the biological synapse.

In addition, Pt/HfO$_x$/ZnO$_x$/TiN device also exhibits the nonlinear transmission efficiency, and the transition from STP to LTP (not shown here) [15].

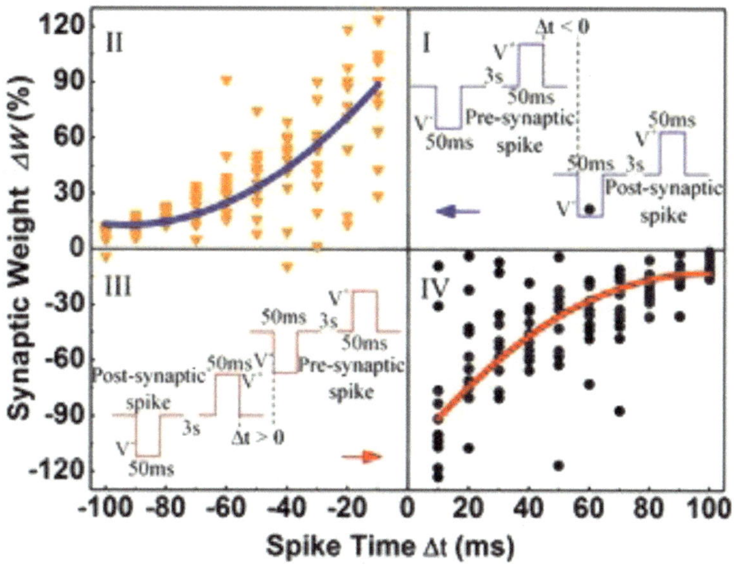

Figure 2.
Emulation of STDP learning rule in Pt/HfO_x/ZnO_x/TiN memristive device—The relative change of the memristor synaptic weight (ΔW) versus the relative spike timing (Δt). And the solid line is the fitting exponential curve to the experimental data. The insets illustrate various spike schemes. The pulse pair comprises a positive and a negative voltage pulse with amplitude of 1.0 V and width of 50 ms. The interval between the two pulses is Δt ms ($t = \pm 10n$, n = 1, 2, ..., 10). The current compliance is not set in the whole emulation process. The current values are read at 0.1 V after 5 min of the spikes [15].

2.2.2 TiN/HfO_x/AlO_x/Pt inorganic memristor

The memristor device based on TiN/HfO_x/AlO_x/Pt can also emulate the biological synapse. Usually, the synapse operates under pulse signals rather than DC bias sweep voltage. The LTPo and LTD phenomena can be observed in Pt/AlO_x/HfO_x/TiN under 180 potentiating pulses (−0.5 V, 10 ms) and 180 depressing pulses (1 V, 10 ms), as shown in **Figure 3a**. The connection strength can be dynamically modulated by the consecutive external signals, determining the transfer efficiency between the electronic neurons.

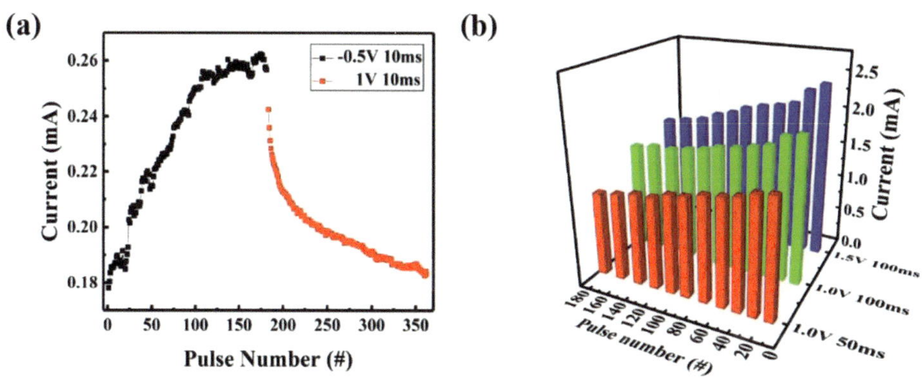

Figure 3.
(a) Change in the response current under the influence of consecutive potentiating or depressing pulses. After pulse stimulation, a 0.1 V 100 ns reading pulse was applied to read the response current. After −0.5 V potentiating pulses, the response current gradually increases (long-term potentiation), while the response current gradually decreases after 1 V depressing pulses (long-term depression). (b) Response of a memristor device to different pulse programs [17].

Further experiments have demonstrated that a pulse signal from amplitude of 1.0–1.5 V and pulse width of 50–100 ms leads to various current responses in **Figure 3b**. That is to say, the larger pulse amplitude, the longer pulse width, and the more pulse number will produce more significant response current change, which is analogous to long-term potentiation/depression of the human brain.

Synaptic plasticity can be divided into STP and LTP according to the timelines of enhanced synaptic connections. The repeated stimulation induced STP to LTP transition is illustrated in **Figure 4**. With increasing the rehearsal pulse number (N = 10, 40, 70, 100, 120), the resistance remaining becomes larger (**Figure 4a**). This procedure is similar to the Ebbinghaus forgetting curve related to human memory [18, 19].

An exponential decay equation was employed to depict the relaxation process:

$$M(t) = M_e + (M_0 - M_e)\ \exp\ (-t/\tau) \tag{2}$$

where $M(t)$, M_0, and M_e are the memory level at time t, t = 0, and at steady state after a long time, and τ of the relaxation time constant. The experimental and simulation results after 70 identical pulses are shown in **Figure 4b**, containing the dependence of τ on N in the inset. The decay rate is faster in the beginning and then becomes slower. The τ value increases from several seconds to 50 seconds with the training pulse number from 10 to 120, revealing a declining forgetting rate from ~57% for N = 10 to ~5% for N = 120. This confirms the transition from STP to LTP through repeated rehearsal and learning.

The STDP rule has been mimicked in $TiN/HfO_x/AlO_x/Pt$ memristor, as indicated in **Figure 5**. The schematic of another training pulse signal with various amplitudes is shown in **Figure 5a**, different from the pulse design in **Figure 2**. A set of pulses (1 V, −0.5 V, −0.45 V, −0.4 V, −0.35 V, −0.3 V, −0.25 V)/(0.5 V, −1 V, −0.9 V, −0.8 V, −0.7 V, −0.6 V, −0.5 V) were used as pre-synaptic/post-synaptic stimulation signals, respectively. Some different pulse signals designed at various spike timings (Δt) are designed and illustrated in **Figure 5b**. When the shortest Δt (10 ms) is inserted to the device, the largest ΔW of 50% for potentiation and − 80% for depression are realized.

The memristive mechanism of asymmetric $TiN/HfO_x/AlO_x/Pt$ memristor has been deeply investigated with the aid of x-ray photoelectron spectroscopy (XPS) depth analyses, which will be discussed in the following Section 2.3.

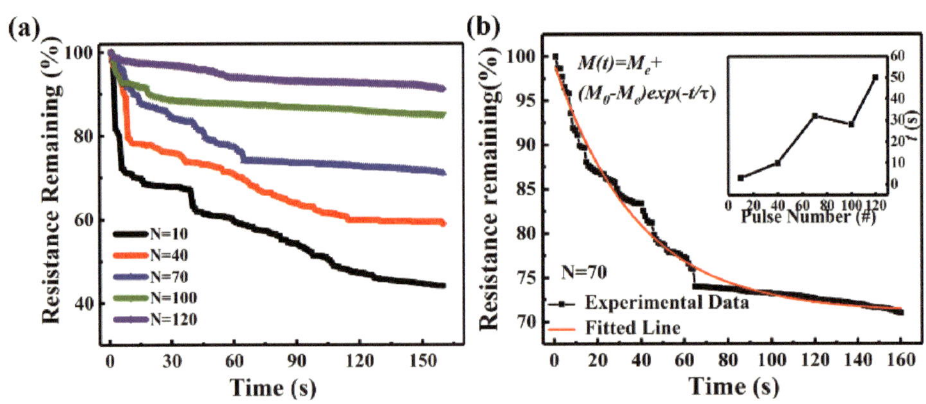

Figure 4.
Repeated stimulation induced STP to LTP transition. (a) Resistance remaining decay curve recorded after 10, 40, 70, 100, 120 identical pulses (1.6 V, 10 ms). A 0.1 V voltage was used to read the device current. (b) Resistance remaining decay curve recorded after 70 identical pulses and the fitted curve according to Eq. (1). The inset plots the dependence of relaxation time τ on the pulse number. τ is obtained by the fitting curve [17].

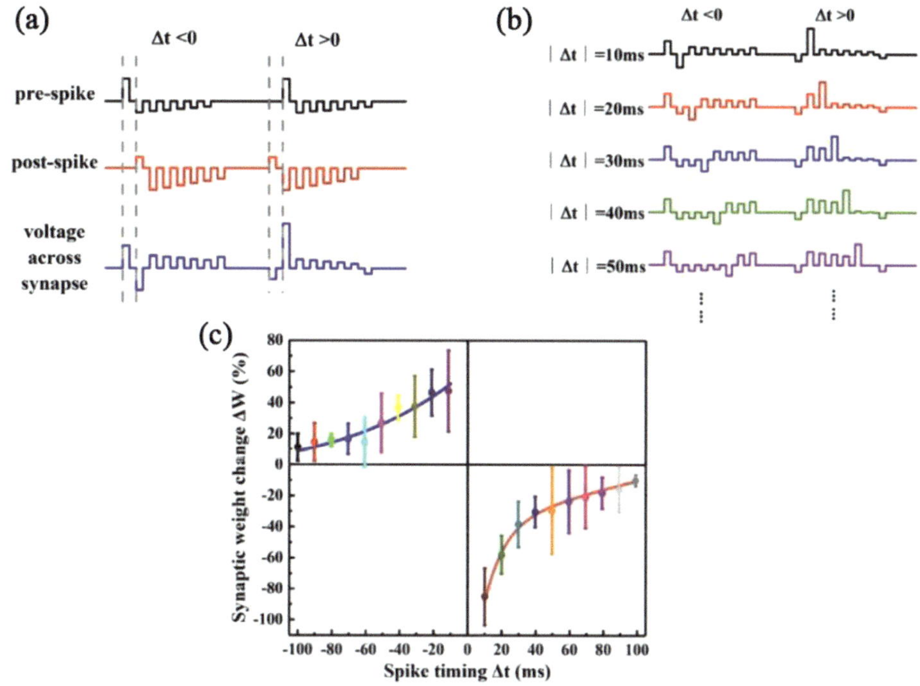

Figure 5.
(a) Training pulse signal with different amplitudes design schematic loaded on the Pt/AlO$_x$/HfO$_x$/TiN memristor (assuming Δt = 10 ms). Each pulse width and pulse interval are 10 ms, respectively. (b) Some different pulse signals designed at various spike timings (Δt). (c) STDP-like curves. The relative change of the memristor synaptic weight (ΔW) versus the relative spike timing (Δt) [17].

2.2.3 TiN/HfO$_x$/HfO$_x$/Pt inorganic memristor

In the previous work on Pt/HfO$_x$/ZnO$_x$/TiN and TiN/HfO$_x$/AlO$_x$/Pt memristors, the asymmetric memristive functional layers of A and B are different materials. Next, we will focus on Pt/HfO$_2$/HfO$_x$/TiN bilayer-structured memristor, as illustrated in **Figure 6a**. 4 nm-thick non-stoichiometric HfO$_x$ films were prepared by

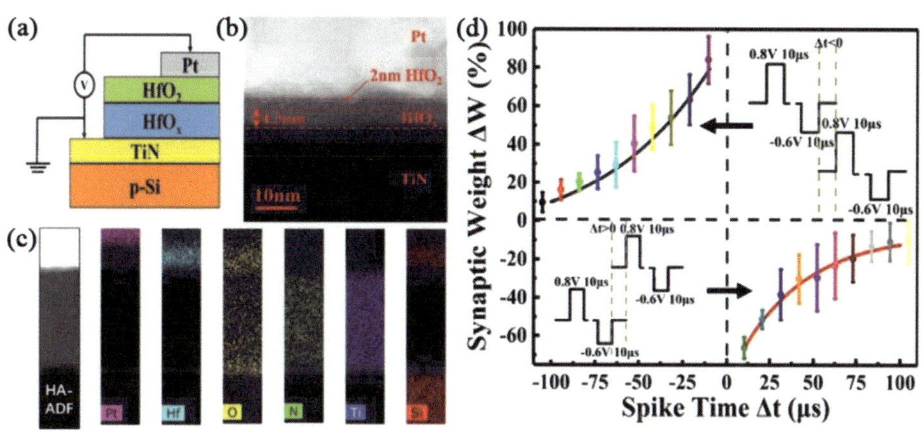

Figure 6.
(a) Schematic of the Pt/HfO$_2$/HfO$_x$/TiN memristor. (b) Cross sectional HAADF-STEM image of the device. (c) EDS elemental mapping of Pt, Hf, O, N, Ti and Si. (d) STDP-like curves. The different synaptic weights (ΔW) versus the different spike times (Δt). The inset shows a pair of pre-synaptic and post-synaptic spikes, and the spike pair is designed to implement STDP [20].

PEALD using the H_2 plasma and 2 nm-thick stoichiometric HfO_2 films by TALD using the H_2O precursor, in basically consistent with the measured result by the cross-sectional high angle annular dark field (HAADF)-scanning transmission microscopy (STEM) in **Figure 6b**. The energy dispersive x-ray spectroscopy (EDS) elemental mapping images of Pt/HfO_2/HfO_x/TiN are shown in **Figure 6c**, revealing the stacking structure. In addition, XPS composition analyses show that the atomic ratio of Hf:O in the HfO_2 and HfO_x layers is 1:2.04 and 1:1.84, respectively, indicating that stoichiometric HfO_2 and nonstoichiometric HfO_x bilayer-structured memristors have been obtained [20]. Hence A and B herein represent HfO_2 and HfO_x with various oxygen contents, respectively. This device unit based on TiN/HfO_x/HfO_2/Pt memristor can also simulate the biological synapse learning rule of STDP, as indicated in **Figure 6d**. When the shortest spike timing of 10 ms is applied to the memristor device, the pulse train responses give rise to the largest ΔW value of 83% for potentiation and − 65% for depression, respectively [20]. These ΔW values for STDP are similar for Pt/HfO_x/ZnO_x/TiN, TiN/HfO_x/AlO_x/Pt and TiN/HfO_x/HfO_2/Pt memristors.

The paired-pulse facilitation (PPF) is a phenomenon wherein the post-synaptic response induced by the spike increases when the time interval of the two spikes is very close [20]. PPF index can be defined as follows:

$$\mathrm{PPF} = (G_2 - G_1)/G_1 - 100\% = C_1 \cdot \exp\left(-\Delta t/\tau_1\right) + C_2 \cdot \exp\left(-\Delta t/\tau_2\right) \quad (3)$$

G_1 and G_2 are the conductance values after the first and the second pulse, respectively. The time constants of τ_1 and τ_2 can be assigned to the fast and slow decaying terms, respectively.

Evidently Pt/HfO_2/HfO_x/TiN memristor displays the marked dependence of synaptic weight on pulse interval Δt by applying the pulse of −1.5 V and 2.5 V, respectively, as seen in **Figure 7a** and **b**. For shortest Δt of 400 ns, the PPF index increases to 135% under positive pulse and becomes −62% under negative pulse. For the negative pulse signals, the calculated τ_1 and τ_2 values are 357 ns and 2.47 ms, respectively; for the positive pulses, τ_1 and τ_2 are 1.48 ms and 6.79 ms, respectively. When the Δt decreases, the memory effect will be improved, which is ascribed to the fact that the smaller Δt between pulses produces less oxygen vacancies to drift back with more effective accumulation of the oxygen vacancies.

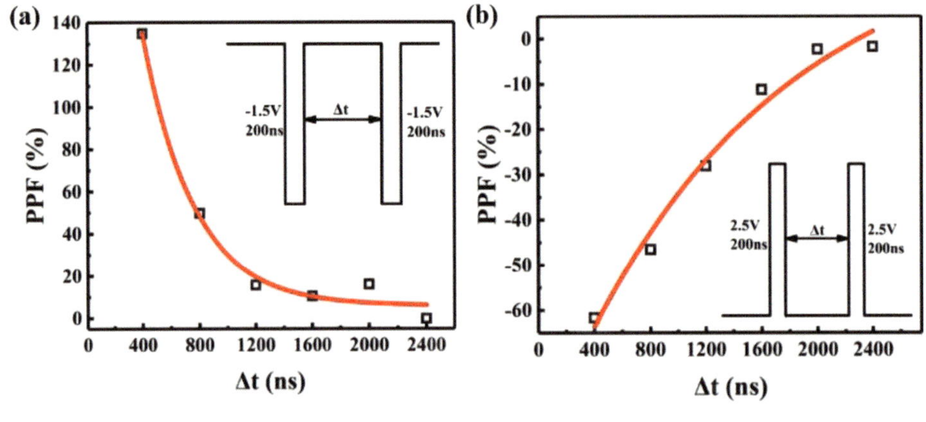

Figure 7.
PPF index as the function of the time interval (Δt) of the Pt/HfO_2/HfO_x/TiN memristor under negative voltage pulse (a) and positive voltage pulse (b). Black points represent the measurement data, and the red lines represent the fitting data by using Eq. (3). The insets in (a) and (b) record the applied pulse waveforms [20].

$Pt/HfO_2/HfO_x/TiN$ also mimics a classical conditioning under different pulse stimuli, as illustrated in **Figure 8**. In the famous experiment [21], a dog salivates (unconditioned response, UR) when watching the food (unconditioned stimulus, US) (**Figure 8a**), when it does not salivate (conditioned response, CR) on hearing the ring (conditioned stimulus, CS) alone (**Figure 8b**). Nonetheless, after some rehearsals, *i.e.* feeding the dog when ringing the bell (**Figure 8c**), the dog salivates even in only hearing the ring (**Figure 8d**). This elucidates that the dog has correlated the food with the ring. Furthermore, when taking away the food, the correlation between the food and ring gradually reduces and even disappears under only the conditioned stimuli (**Figure 8e**). The whole procedure can be emulated in the $Pt/HfO_2/HfO_x/TiN$ by using +4 V and − 1.3 V stimuli with a single pulse duration of 5 ms.

Before rehearsing, the memristor has a low resistance state of 5 kΩ. The +4 V stimulus (US) causes a high resistance state of 3 MΩ (UR) (**Figure 8f**), when the −1.3 V stimulus (CS) only results in a low resistance state of 5 kΩ before training (**Figure 8g**). In Pavlov's experiments, the food and the ring exist simultaneously to reinforce the correlation between US and CS. In our experiments, the +2.7 V stimulus was exerted to the memristor, the same as the simultaneous stimuli of −1.3 V and + 4 V pulse signals. When two rehearsing sequences with +2.7 V pulse, the device becomes the high resistance output of 2 MΩ (CR) (**Figure 8h**). When removing the +4 V signal, the memristor continues to keep in a high resistance state under a series of −1.3 V stimuli alone and then returns to a low resistance state (**Figure 8i**), implying the setup and vanish of the classical conditional reflex.

The energy consumption is one important indicator for a practical electronic synaptic device in neuromorphic network. $Pt/HfO_2/HfO_x/TiN$ memristor can be set in less than 100 ns and reset in less than 10 ns, indicating the rapid switching speed, as recorded in **Figure 9a**.

The current response curves versus the time after the applied programming signal during the set or reset operation are plotted in **Figure 9b** and **c**, respectively. The current rises after a waiting time of about 260 ns when a − 2 V/1 ms stimulus is applied, indicating the beginning of the set process (**Figure 9b**). The memristor resistance decreases from the initial high resistance state (~1 MΩ) to low resistance state (~800 Ω). Similarly, the current reduces after a waiting time of about 70 ns when a +3 V/1 ms signal is exerted, showing the occurrence of the reset process. The energy consumption per operation can be calculated to be 520 pJ for the set process and 1.05 nJ for the reset process by considering the pulse waveforms (time, response current, and pulse voltage), corresponding to the maximum energy consumption in one set or reset operation, as the memristor has been set in the lowest resistance state with the highest response current. Nevertheless, the actual operation of the electronic synapse is generally in the mediate resistance states (~80 kΩ). The response current of the memristor is inversely proportional to the resistance value of the synaptic device with a first-order approximation. So, the evaluated actual energy consumption per operation will decline in the range of around ten picojoules.

Finally, the impact of oxygen vacancy concentration in non-stoichiometric HfO_x layers on resistive switching properties of $Pt/HfO_2/HfO_x/TiN$ bilayer ultrathin memristor has been investigated. The memristor with 12.1% oxygen vacancy concentration in the HfO_x layer exhibits comprehensively better performances such as the optimal pulse energy consumption, reset switching speed, and DC endurance and retention characteristics [20].

2.2.4 TaN/Ti-MA/TiO_2/Pt organic: inorganic hybrid memristor

As mentioned above, we mainly elucidated the bio-synaptic functions of three asymmetric inorganic bilayer-structured ultrathin memristors. In this part,

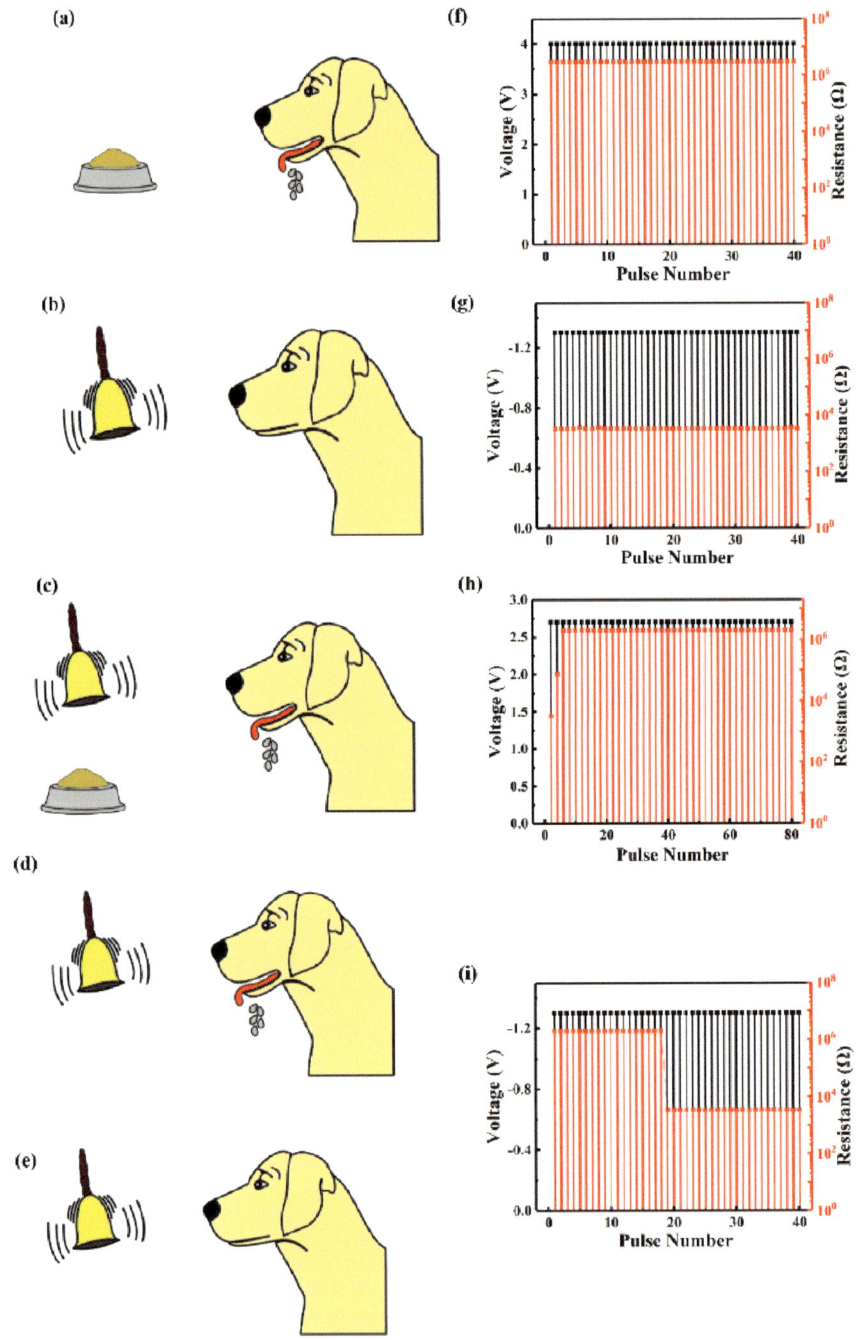

Figure 8.
Emulation of the acquisition and extinction of classical conditioning demonstrated by Pavlov's dog experiment. (a) The dog salivates while watching the food (US → salivating UR). (b) The dog does not salivate upon hearing the ringing alone (CS → no salivating). (c) In the training process, the food and ringing together stimulates the dog, and the dog salivates (US + CS → salivating). (d) After sufficient training, classical conditioning is formed and the dog salivates upon hearing the ringing alone (CS → salivating, CR). (e) Extinction of classical conditioning after removing the food for some time. (f) Positive +4 V pulse (US) can lead to a high resistance output, similar to the UR in (a). (g) The negative −1.3 V pulse (CS) cannot lead to the high resistance output before training, similar to the CS in (b). (h) After applying several training sequences of +2.7 V pulse voltage, equal to the simultaneous stimuli of −1.3 V and +4 V pulses, the device reaches a high resistance state of 2 MΩ (CR), analogous to the phenomenon in (c). (i) After only applying some −1.3 V pulse alone, the memristor remains in a high resistance state, similar to CR in (d). Then it returns to the low resistance state, consistent with the extinction of classical conditioning in (e) [20].

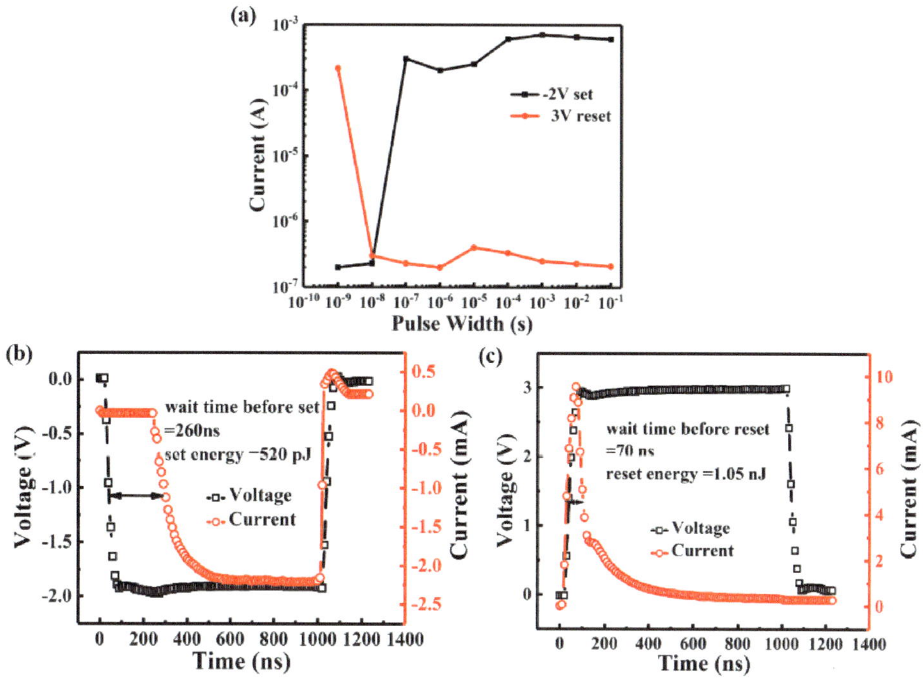

Figure 9.
(a) Switching speed test of the synaptic device. The voltage for the set and reset operation in the memristor is about −2 V and +3 V, respectively. The device can be switched in less than 100 ns for a set operation and less than 10 ns for a reset operation. (b) Transient current response on the applied voltage pulse for a set operation from the high resistance state to the low resistance state. The set pulse amplitude, width, rising time, and falling time are set to be −2 V, 1000 ns, 20 ns, and 20 ns, respectively. (c) Transient current response on the applied voltage pulse for a reset operation from the low resistance state to the high resistance state. The reset pulse amplitude, width, rising time, and falling time are set to be 3 V, 1000 ns, 20 ns, and 20 ns, respectively [20].

organic–inorganic hybrid bilayer memristors of TaN/Ti-MA/TiO_2/Pt were prepared by low temperature MLD/ALD at 160°C. The synaptic plasticity has been explored deeply. Some superb synaptic functions, such as nonlinear transmission characteristics, STP/LTP, PPF, and STDP have been achieved in the hybrid memristors [22].

First the narrow-scan XPS and Fourier transform infrared (FTIR) spectroscopy were used to detect the chemical composition and organic group of Ti-based maleic acid (Ti-MA) hybrid film, as shown in **Figure 10a–d**. The C 1 s XPS peaks at 284.6 eV and 288.4 eV (**Figure 10a**) result from the C-C (backbone chain carbon) bond and the O-C=O bond from carboxyl, respectively, suggesting the occurrence of organic component in Ti-MA films. The doublet at 458.7 eV and 464.5 eV with the spin orbit splitting energy of 5.8 eV can be assigned to the Ti $2p_{1/2}$ and Ti $2p_{3/2}$ ones from the Ti-O bond of TiO_2 [13, 23] (**Figure 10b**, which indicates the inorganic component in hybrid films. Moreover, the O 1 s spectrum can be deconvoluted into two peaks at 530.0 eV and 531.6 eV, corresponding the O-Ti and O-C bonds, respectively (**Figure 10c**). The FTIR spectrum of Ti-MA hybrid film (**Figure 10d**) displays the asymmetric and symmetric stretch of carboxylate groups at 1575 cm^{-1} and 1447 cm^{-1}. The splitting of 128 cm^{-1} indicates the bidentate bond mode between the Ti ion and carboxyl. As a result, Ti-MA inorganic–organic hybrid films have been fabricated successfully.

The resistive switching characteristics of the hybrid bilayer memristor of TaN/Ti-MA/TiO_2/Pt have been examined for 100 times, as seen in in **Figure 11a**. The typical bipolar resistive switching behavior has been confirmed with narrow distribution of set voltage of -1.6 ± 0.2 V (red line) or reset voltage of 1 ± 0.1 V

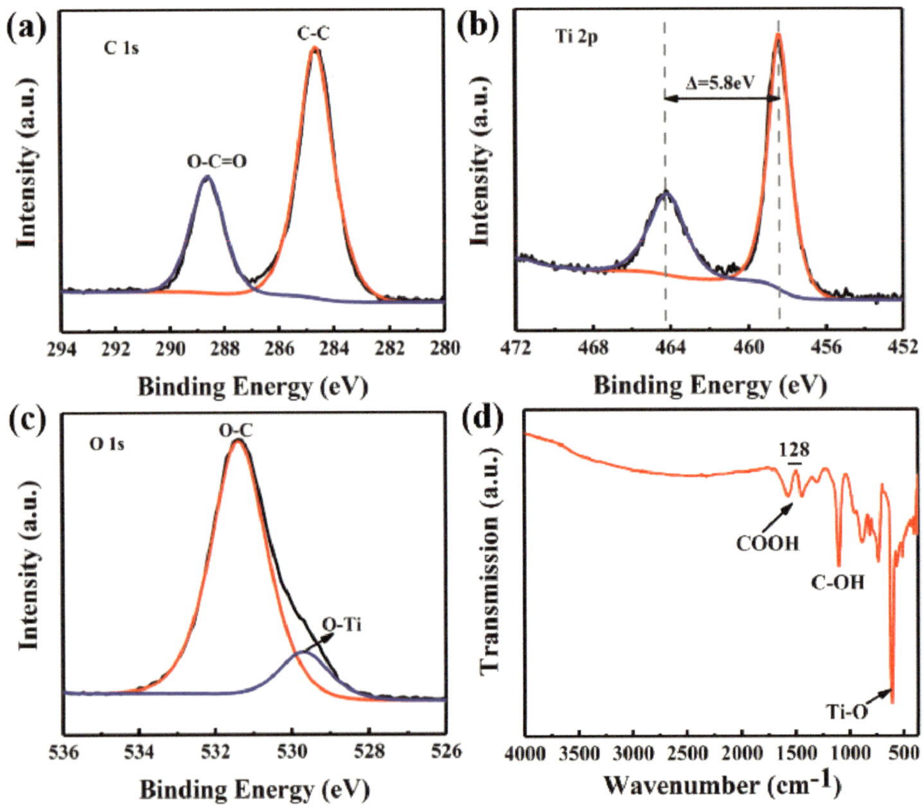

Figure 10.
Narrow-scan XPS spectra of (a) C 1 s, (b) Ti 2p and (c) O 1 s and (d) FTIR spectrum from the Ti-MA hybrid films on Si [22].

(black line). The double-logarithmic *I-V* curves and linear fits to the set process are shown in **Figure 11b**. At the low voltage stage, the I-V is dominated by the Ohm's law with the approximately linear relationship (region 1, R^2 = 0.9996). When the voltage increases, the current is dependent of near square of the voltage, obeying the Child conductive law (region 2, R^2 = 0.9995). At critical voltage of around 1.2 V, the current is proportional to the *n*th power of the voltage with a sharp current rise

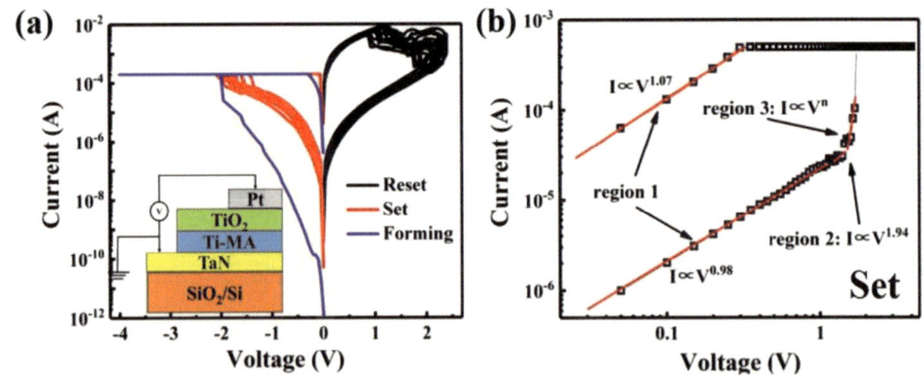

Figure 11.
(a) I-V curve of the TaN/Ti-MA/TiO_2/Pt hybrid memristor for 100 times DC ramp voltages tests. Bottom inset is the schematic of the memristor. (b) Double-logarithmic I-V curves and linear fits to the set process [22].

(region 3, R^2 = 0.9904). All these prove the space charge limited current (SCLC) model in hybrid bilayer memristor [24], revealing the filament model of oxygen vacancy migration.

The PPF and STDP functions have also been characterized in hybrid memristor, as shown in **Figures 12** and **13**, respectively. A pair of pulses (−1 V, 400 ns) with different Δt were applied to the hybrid memristor (**Figure 12a**). The measured data can be well fit exponentially (**Figure 12b**). The PPF index has reached to 361% with the 400 ns pulse interval in hybrid memristor. When the pulse interval increases to 2400 ns, the PPF index dramatically tends to 3% [22]. Compared to inorganic bilayer memristor, the organic–inorganic hybrid bilayer device has much larger PPF index in the same pulse interval of 400 ns.

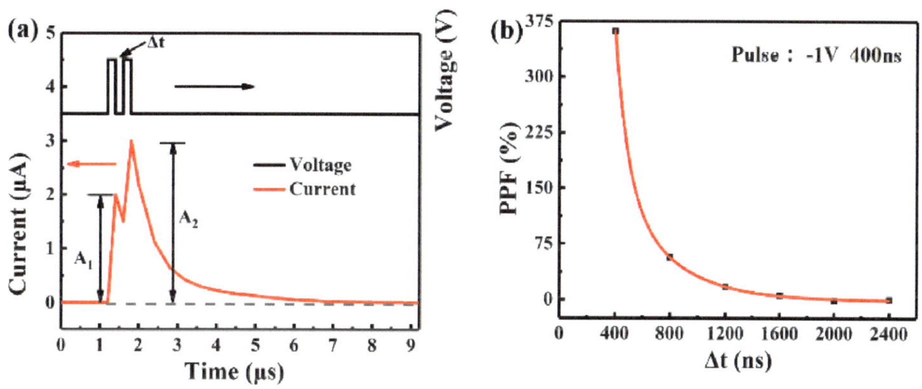

Figure 12.
(a) PPF function in memristors generated by two pulse spikes and the real-time response current. (b) PPF curves with different pulse interval time [22].

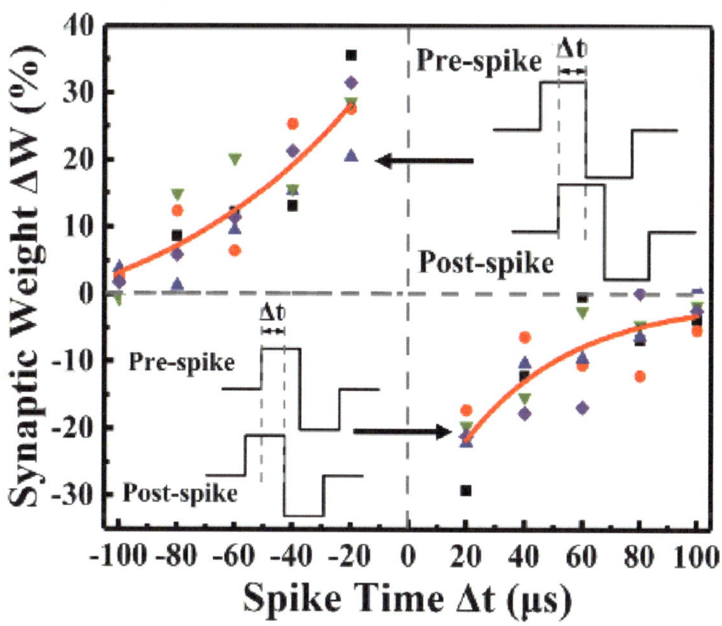

Figure 13.
STDP curves obtained in hybrid memristor. The spot is the measured data and the red line is the fitting results. The insets are the spike pulse signals designed by a pair of 0.8 V and −0.8 V pulses with pulse width of 120 μs [22].

The STDP rule was emulated in hybrid memristor by using a pair of 0.8 V and −0.8 V pulses with 120 μs pulse width. The ΔW has a strong time correlation with maximum 35% increment at the Δt of −20 μs and −20% reduction at the Δt of 20 μs. These values are relatively smaller than the ΔW maximum value of 60–90% of inorganic memristors. Finally, the ΔW in hybrid device obeys the exponential association with the Δt, namely

$$\Delta W = A\ \exp\ (-\Delta t/\tau) \tag{4}$$

The measured data can be fitted well.

In addition, the conditioned reflex has been mimicked in hybrid film memristor, similar to the results of Pt/HfO_2/HfO_x/TiN memristor in **Figure 8**.

By comparison with inorganic bilayer memristors, it can be found that the organic–inorganic hybrid bilayer memristor has similar bio-synaptic functions with comparable switching speed and energy consumption. Moreover organic–inorganic hybrid materials may possess both the advantages of organic and inorganic components with excellent flexibility and tunability. Inorganic compounds have better electrical characteristics and thermal stability. Organic compounds own various functional groups, larger stretchability and low processing temperature. By means of the synergetic and complementary effects between organic and inorganic components, the comprehensive properties of hybrid memristive materials could be expected for significant improvement. The hybrid bilayer ultrathin memristor derived by low temperature MLD/ALD is one competitive candidate for flexible neuroscience applications.

2.3 Memristive mechanism

In Section 2.2, we focused on the electrical Performance and synaptic functions of several bilayer ultrathin memristors. In this section, the asymmetric memristive mechanism of the bilayer-structured memristors on TiN or TaN will be studied carefully. Taking Pt/AlO_x/HfO_x/TiN memristor as an example, the XPS depth profiles of asymmetric bilayer device units were obtained under various resistance states of the initial state, low resistance state (LRS), high resistance state (HRS), and medium resistance state [17]. XPS is a powerful surface analytical tool to determine the chemical valence and the oxygen vacancy contents in multilayer-structured metal oxide thin films [23, 25].

Figure 14a–d records the high-resolution Al 2p, Hf 4f and O 1 s peaks in AlO_x and HfO_x layers for as-deposited Pt/AlO_x/HfO_x/TiN in the initial state. The Hf 4f spectra from the HfO_x layer can be deconvoluted into four peaks (**Figure 14b**). The stronger peaks at ~16.7 eV and 18.6 eV originate from the Hf^{4+} in the HfO_x layer, whereas the weaker ones with slightly lower energies of 15.6 eV and 17.9 eV come from the $Hf^{(4-x)+}$ in the low valence Hf sub-oxide. The content percentage of two Hf valence states in the HfO_x layer can be roughly evaluated by calculating the area proportion of each peak, as shown in the inset of **Figure 14b** [26–28]. The percentage of Hf^{4+} and $Hf^{(4-x)+}$ in the HfO_x layer is around 89.7% and 9.4%, respectively. A similar analysis can be also carried out for the Al 2p spectra from AlO_x layer (**Figure 14a**). Meanwhile the O 1 s spectra from the AlO_x and HfO_x layers can also be deconvoluted into two peaks. The stronger peaks at around 531.5 and 531.0 eV result from Al-O and Hf-O bonding in the AlO_x and HfO_x layers, respectively, whereas the weaker ones with a slightly higher energy of 532.1 eV in the O 1 s spectra are ascribed to the oxygen vacancies in the AlO_x and HfO_x layers according to the literature reports [26–29]. The calculated percentage of oxygen vacancies in the AlO_x and HfO_x layers is around 0.7% and 8.1%, respectively (**Figure 14c** and **d**).

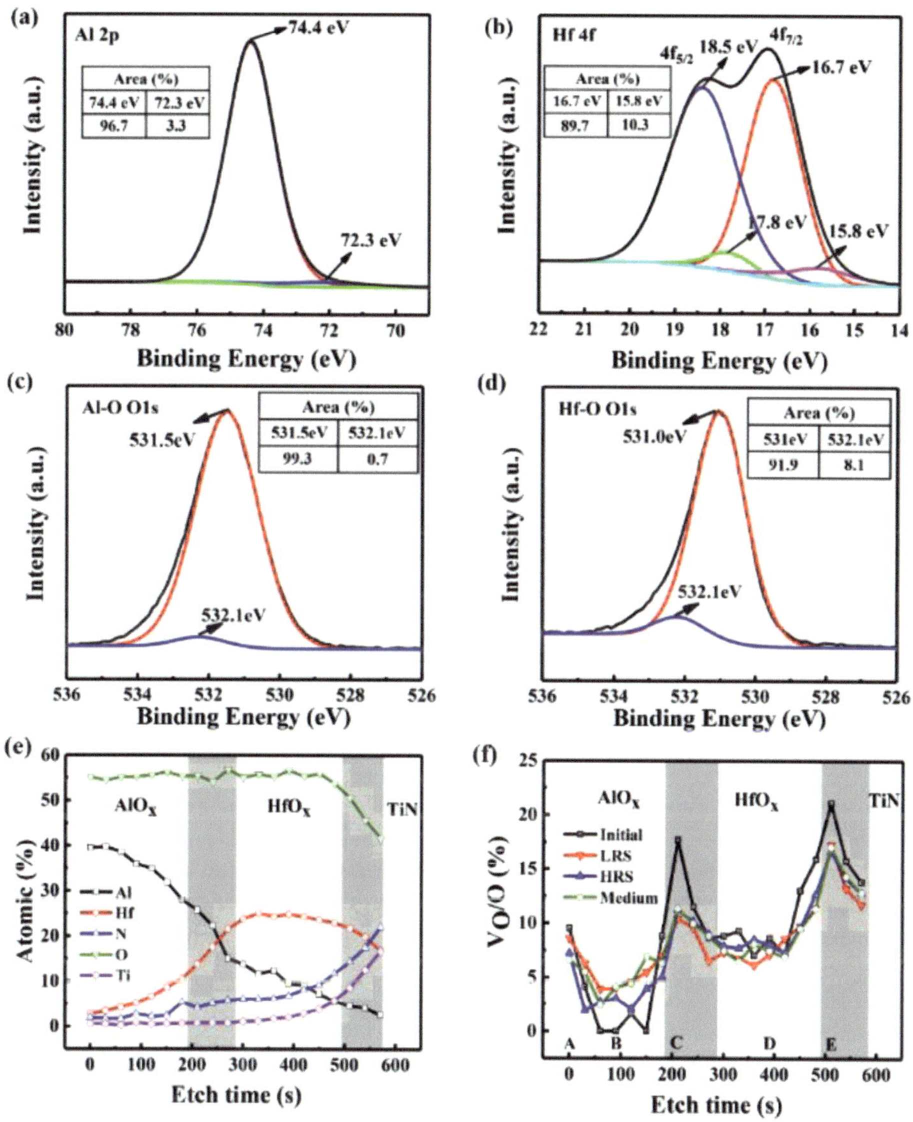

Figure 14.
Narrow-scan (a) Al 2p, (b) Hf 4f and O 1 s peaks of (c) AlO_x and (d) HfO_x in as-prepared Pt/AlO_x/HfO_x/TiN in the initial state. (e) XPS depth profile of Pt/AlO_x/HfO_x/TiN in the initial state. (f) The depth distribution of the average oxygen vacancy concentration in the initial state, LRS, HRS, and medium resistance state after 40 pulses (1.5 V, 0.5 ms). The gray region in figure (e) and (f) is the interfacial layer [17].

Significantly, the oxygen vacancy content in the HfO_x layer is much higher than that in the AlO_x layer.

The XPS depth data of Pt/AlO_x/HfO_x/TiN memristor by Ar ion etching under various resistance states may provide some valuable information on the valence states and defects of metal oxide layers [24], for the initial state sample recorded in **Figure 14e**. The AlO_x/HfO_x bilayer structure could be recognized with an evident interfacial diffusion between AlO_x/HfO_x and HfO_x/TiN (gray region). The depth distribution of the average oxygen vacancy concentration in the asymmetric Pt/AlO_x/HfO_x/TiN memristors under various resistance states of the initial state, LRS, HRS, and medium resistance state is illustrated in **Figure 14f**. **Table 3** lists the average oxygen vacancy concentration values of Pt/AlO_x/HfO_x/TiN for four resistance states at different positions of A, B, C, D, and E, corresponding to an etch time of 0 s, 90 s, 210 s, 390 s, and 510 s. Herein A, B, C, D, and E locate in the interface of the Pt/AlO_x, AlO_x layer, the interface of the AlO_x/HfO_x, HfO_x layer, and the interface of HfO_x/TiN, respectively. The oxygen vacancy distribution is inhomogeneous in the Pt/AlO_x/HfO_x/TiN memristor, and the oxygen vacancy concentration of the interfaces between AlO_x/Pt (A), AlO_x/HfO_x (C), HfO_x/TiN (E) is markedly higher than that of the adjacent AlO_x (B) and HfO_x (D) layers. Furthermore, the oxygen vacancy concentration in HfO_x (D) is much higher than that in AlO_x (B) layer.

In general, the resistive switching mechanism of metal oxide memristors is related to the connection and rupture of conductive filaments of oxygen vacancies. But the simple increase of oxygen vacancy concentration is not always effective. The non-uniform distribution of oxygen vacancies in memristors is the critical factor affecting the resistive switching behavior of memristive devices [30].

Based on the oxygen vacancy concentration and distribution in the Pt/AlO_x/HfO_x/TiN memristors under various resistance states in **Figure 14f**, we proposed a memristive mechanism of an asymmetric bilayer metal oxide synaptic device to explain synaptic plasticity, as illustrated in **Figure 15**.

There are much more random oxygen vacancies in the HfO_x layer than in the AlO_x layer for as-deposited Pt/AlO_x/HfO_x/TiN device. Meanwhile, the oxygen vacancy concentration in the interfaces of AlO_x/HfO_x and HfO_x/TiN is evidently higher than the HfO_x layer (**Figure 15a**). During the forming process, the disorderly distributed oxygen vacancies in the bilayer oxide layers and interfacial layers form conductive filaments under the external electrical field, similar to the soft breakdown of the capacitor. So the connection and disconnection of the conductive filaments lead to resistive switching. When inserting a −3 V forming voltage, the device turns from the initial state to LRS with suddenly resistance drop from 10 MΩ to 600 Ω, suggesting that the oxygen vacancies with positive charges (V_O^{2+}) in the AlO_x/HfO_x interface, HfO_x layer, and HfO_x/TiN interface move to the AlO_x layer and AlO_x/Pt interface. Simultaneously, the oxygen vacancy concentration gradient help to the migration of the oxygen vacancies, forming localized conductive filaments of oxygen vacancies in the bilayer structured AlO_x/HfO_x device (**Figure 15b**).

After applying the +2.5 V reset voltage to the LRS device, the memristor transfers from LRS (600 Ω) to HRS (1 MΩ) (**Figure 15c**). During the reset process, the oxygen vacancies migrate from the AlO_x/Pt interface and AlO_x layer to the AlO_x/HfO_x interface and HfO_x layer, leading to the rupture of oxygen vacancy conductive filaments in the AlO_x layer. Besides, considering the thermophoresis/diffusion-driven oxygen migration [31, 32], the middle position of the conductive filament in the AlO_x layer first breaks up, causing a spatial gap, as indicated by the red arrow in **Figure 15c**. *electron* tunneling happens through the physical gap with the enhanced resistance. During the reset process, the oxygen vacancy concentration declines at the AlO_x/Pt interface and AlO_x layer and rises at the AlO_x/HfO_x interface and HfO_x

Oxygen vacancy concentration	**A**	**B**	**C**	**D**	**E**
Position	Pt/AlO_x	AlO_x	AlO_x/HfO_x	HfO_x	HfO_x/TiN
Initial	9.5%	0.7%	17.7%	8.7%	21.1%
LRS	8.8%	4.1%	10.4%	7.1%	17.3%
HRS	7.2%	3.0%	11.2%	8.0%	16.7%
Medium	6.6%	4.0%	11.3%	7.5%	17.0%

Table 3.
Average oxygen vacancy concentration of Pt/AlO_x/HfO_x/TiN in various positions for different resistance states [17].

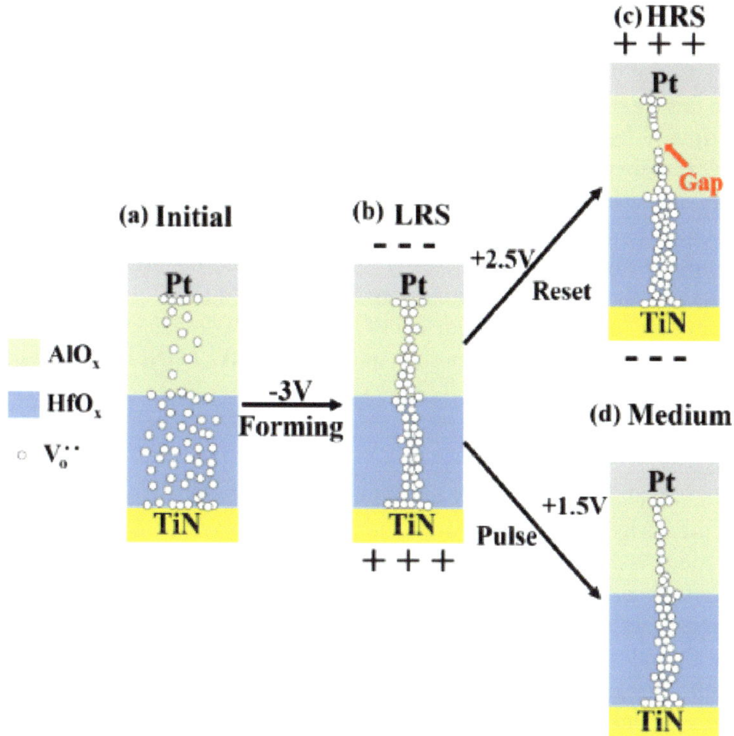

Figure 15.
Model of the formation and rupture of a conductive filament consisting of oxygen vacancies. After −3 V forming voltage, the device transfers from an initial resistance state (a) to a low resistance state (LRS) (b); after 2.5 V reset voltage, the device transfers from a low resistance state to a high resistance state (HRS) (c); after 40 continuous pulses (+1.5 V, 10 ms), the device transfers from LRS to a medium resistance state (d) [17].

layer, as proved by the relative variation of oxygen vacancy concentrations in **Figure 14f** and **Table 3**.

Because the synapse device usually operates under pulse mode to get more intermediate resistance states, a continuous pulse experiment was exploited to alter the device from a LRS (600 Ω) to a medium resistance state (50 kΩ) by imposing 40 pulses (+1.5 V, 10 ms) (**Figure 15d**). The XPS result (**Figure 14f**) indicates that the oxygen vacancy concentration curve of medium resistance state devices lies approximately between HRS and LRS in the AlO_x layer and HfO_x layer. The difference in oxygen vacancy concentration at the interface layers of AlO_x/HfO_x and HfO_x/TiN among the medium resistance state, LRS, and HRS is slightly little. In consequence, during regular operations of synaptic memristors, the formation/rupture of nanoscale conductive filaments tend to appear in the low *k* AlO_x layer with lower electric field intensity [33]. Furthermore, the device conductance can be modulated by the oxygen vacancy drift under pulse electric field, producing a change in the concentration and distribution of oxygen vacancies at the interface of the metal/oxide and the interior. In the LRS of 600 Ω, the conductive filament is thick with the conductance of 22 G_0, corresponding to a wide conductive filament with classical metallic properties. After 40 pulse stimuli, a medium resistance state of 50 kΩ is obtained with a conductance of 0.26 G_0, where the conductive filament behaves as a quantum wire, producing a single-defect conducting path [31, 33, 34].

The oxygen vacancy migration/diffusion model can be used to explain the transition from STP to LTP in bilayer memristive device (**Figure 4a**). When imposing the +1.6 V pulse, the oxygen vacancies move from the AlO_x layer to the HfO_x layer with the reduced response current. When the voltage is removed, some oxygen vacancies may stay in a new steady position, however some oxygen vacancies may diffuse back to the old position owing to the gradient of oxygen concentration. This leads to the device conductance change with a reduced synaptic weight during the relaxation time. After applying repetitive pulse stimuli, the subsequent voltage forces the reversely diffused oxygen vacancies to move forward again so as to improve the migration efficiency until most oxygen vacancies attain new equilibrium positions. The remaining synaptic weight gradually increases with the increasing pulse number. This process is called repeated training and learning, corresponding to the transformation from STP to LTP [17].

The memristive mechanism from $Pt/AlO_x/HfO_x/TiN$ device is also applicable to other bilayer-structured memristors such as $Pt/HfO_2/HfO_x/TiN$, $Pt/TiO_2/Ti\text{-}MA/TaN$.

3. Conclusion

Our asymmetric bilayer-structured memristors fabricated by ALD/MLD and their main memristive features are summarized in **Table 4**, including set/reset voltage, ON/OFF ratio, and some important synaptic functions. Some similar work with asymmetric bilayer structure has also been listed in **Table 4** for comparison. It can be seen that all memristors with asymmetric bilayer structure exhibit better resistive switching performance. Our memristors have relatively thinner functional layers, relatively smaller ON/OFF ratio and emulate more artificial synaptic functions such as LTPo, LTD, the transition from STP to LTP, PPF, STDP, and conditional reflex (CR). The memristive mechanism of our bilayer-structured ultrathin device has been proposed to explain the synaptic plasticity based on oxygen vacancies migration/diffusion model. The non-uniform distribution of oxygen vacancies in asymmetric bilayer memristors plays the crucial role in affecting the linkage/rupture of conductive filaments.

In light of these promising results and the fabrication compatibility with semiconductor industry, the ALD/MLD-derived bi-layer ultrathin memristor devices have tremendous potential as billions of electronic synapses in next-generation artificial neural network and flexible electronics.

Device structure	Thickness (nm)	Set/reset voltage (V)	ON/OFF ratio	Synaptic functions	References
$Pt/HfO_2/ZnO/TiN$	~10	−1.7/+1.4	~30	LTPo, LTD, STP/LTP, STDP	Our works
$Pt/Al_2O_3/HfO_2/TiN$	~10	−1.4/ +1.3	~610	LTPo, LTD, STP/LTP, PPF, STDP	
$Pt/HfO_2/HfO_x/TiN$	~6	−1.6/+1.1	~954	LTPo, LTD, STP/LTP, PPF, STDP, CR	
$Pt/TiO_2/Ti\text{-}MA/TaN$	~8	−1.5/+1	~230	LTPo, LTD, STP/LTP, PPF, STDP, CR	
$Ni/SiN_x/AlO_y/TiN$	~11.5	+4/−3.5	~500	LTPo, LTD, STDP	[35]
$TiN/HfO_2/Al_2O_3/Pt$	~10	+1.4/−1.3	$\sim10^5$	STDP	[36]
$W/AlO_x/Al_2O_3/TiN$	~10	+1.05/−1.25	$\sim10^3$	—	[37]
$Ag/ZrO_2/WS_2/Pt$	~100	+0.16/−0.06	$>10^5$	PPF, STDP	[38]

Table 4.
Comparison of main memristive features of our ultrathin memristors fabricated by ALD/MLD and other asymmetric bilayer-structured memristors.

Acknowledgements

This work is supported by the Natural Science Foundation of China (52073142, 51721001) and Jiangsu Province (BK20201252). Mr. Chang Liu also acknowledges the support from the Nanjing University Innovation and Creative Program for PhD candidate (CXCY18-27).

Conflict of interest

The authors declare no conflict of interest.

Author details

Chang Liu, Lin Zhu, Lai-Guo Wang and Ai-Dong Li*
National Laboratory of Solid State Microstructures, Department of Materials Science and Engineering, College of Engineering and Applied Sciences, Jiangsu Key Laboratory of Artificial Functional Materials, Collaborative Innovation Center of Advanced Microstructures, Nanjing University, Nanjing, People's Republic of China

*Address all correspondence to: adli@nju.edu.cn

References

[1] Chua L. Memristor-the missing circuit element. IEEE Transactions on circuit theory. 1971;18(5):507-519.

[2] Strukov DB, Snider GS, Stewart DR, Williams RS. The missing memristor found. nature. 2008;453(7191):80-83.

[3] Midya R, Wang Z, Zhang J, Savel'ev SE, Li C, Rao M, et al. Anatomy of Ag/Hafnia-based selectors with 10^{10} nonlinearity. Advanced Materials. 2017; 29(12):1604457.

[4] Wang Z, Joshi S, Savel'ev SE, Jiang H, Midya R, Lin P, et al. Memristors with diffusive dynamics as synaptic emulators for neuromorphic computing. Nature materials. 2017;16 (1):101-108.

[5] Wang J, Pan R, Cao H, Wang Y, Liang L, Zhang H, et al. Anomalous rectification in a purely electronic memristor. Applied Physics Letters. 2016;109(14):143505.

[6] Du C, Ma W, Chang T, Sheridan P, Lu WD. Biorealistic implementation of synaptic functions with oxide memristors through internal ionic dynamics. Advanced Functional Materials. 2015;25(27):4290-4299.

[7] Azghadi MR, Chen YC, Eshraghian JK, Chen J, Chang YF. Complementary Metal oxide Semiconductor and Memristive Hardware for Neuromorphic Computing. Advanced Intelligent Systems. 2020:1900189.

[8] Kim S, Chen J, Chen YC, Kim MH, Kim H, Kwon M, et al. Neuronal dynamics in HfO_x /AlO_y -based homeothermic synaptic memristors with low-power and homogeneous resistive switching. Nanoscale. 2019;11(1):237-245.

[9] Puurunen RL. Surface chemistry of atomic layer deposition: A case study for the trimethylaluminum/water process. Journal of applied physics. 2005;97(12):9.

[10] Sherman A. Atomic layer deposition for nanotechnology: an enabling process for nanotechnology fabrication: Ivoryton Press; 2009.

[11] Arden WM. The international technology roadmap for semiconductors —perspectives and challenges for the next 15 years. Current Opinion in Solid State and Materials Science. 2002;6(5): 371-377.

[12] Kot M, Das C, Wang Z, Henkel K, Rouissi Z, Wojciechowski K, et al. Room-temperature atomic layer deposition of Al_2O_3: Impact on efficiency, stability and surface properties in perovskite solar cells. ChemSusChem. 2016;9(24).

[13] Cao Y-Q, Zhang W, Xu L, Liu C, Zhu L, Wang L-G, et al. Growth mechanism, ambient stability, and charge trapping ability of Ti-Based maleic acid hybrid films by molecular layer deposition. Langmuir. 2019;35(8): 3020-3030.

[14] Yu S, Gao B, Fang Z, Yu H, Kang J, Wong HSP. A low energy oxide-based electronic synaptic device for neuromorphic visual systems with tolerance to device variation. Advanced Materials. 2013;25(12):1774-1779.

[15] Wang L-G, Zhang W, Chen Y, Cao Y-Q, Li A-D, Wu D. Synaptic plasticity and learning behaviors mimicked in single inorganic synapses of Pt/HfO_x/ZnO_x/TiN memristive system. Nanoscale research letters. 2017;12(1): 1-8.

[16] Mori M, Abegg MH, Gähwiler BH, Gerber U. A frequency-dependent switch from inhibition to excitation in a hippocampal unitary circuit. Nature. 2004;431(7007):453-456.

[17] Liu C, Wang L-G, Cao Y-Q, Wu M-Z, Xia Y-D, Wu D, et al. Synaptic functions and a memristive mechanism on Pt/AlO_x/HfO_x/TiN bilayer-structure memristors. Journal of Physics D: Applied Physics. 2019;53(3):035302.

[18] Wang ZQ, Xu HY, Li XH, Yu H, Liu YC, Zhu XJ. Synaptic learning and memory functions achieved using oxygen ion migration/diffusion in an amorphous InGaZnO memristor. Advanced Functional Materials. 2012;22 (13):2759-2765.

[19] Jo SH, Chang T, Ebong I, Bhadviya BB, Mazumder P, Lu W. Nanoscale memristor device as synapse in neuromorphic systems. Nano letters. 2010;10(4):1297-1301.

[20] Liu C, Zhang C-C, Cao Y-Q, Wu D, Wang P, Li A-D. Optimization of oxygen vacancy concentration in HfO_2/HfO_x bilayer-structured ultrathin memristors by atomic layer deposition and their biological synaptic behavior. Journal of Materials Chemistry C. 2020; 8(36):12478-12484.

[21] Wu C, Kim TW, Guo T, Li F, Lee DU, Yang JJ. Mimicking classical conditioning based on a single flexible memristor. Advanced Materials. 2017;29 (10):1602890.

[22] Liu C, Cao Y-Q, Wu D, Li A-D. Simulation of Biologic Synapse Through Organic-Inorganic Hybrid Memristors Using Novel Ti-Based Maleic Acid/TiO_2 Ultrathin Films. IEEE Electron Device Letters. 2019;41(1):155-158.

[23] Zhang W, Kong J-Z, Cao Z-Y, Li A-D, Wang L-G, Zhu L, et al. Bipolar resistive switching characteristics of HfO_2/TiO_2/HfO_2 trilayer-structure RRAM devices on Pt and TiN-coated substrates fabricated by atomic layer deposition. Nanoscale research letters. 2017;12(1):1-11.

[24] Pan F, Gao S, Chen C, Song C, Zeng F. Recent progress in resistive random access memories: materials, switching mechanisms, and performance. Materials Science and Engineering: R: Reports. 2014;83:1-59.

[25] Hu L, Zhu S, Wei Q, Chen Y, Yin J, Xia Y, et al. Enhancement of resistive switching ratio induced by competing interfacial oxygen diffusion in tantalum oxide based memories with metal nitride electrode. Applied Physics Letters. 2018;113(4):043503.

[26] Huang C-Y, Huang C-Y, Tsai T-L, Lin C-A, Tseng T-Y. Switching mechanism of double forming process phenomenon in ZrO_x/HfO_y bilayer resistive switching memory structure with large endurance. Applied Physics Letters. 2014;104(6):062901.

[27] Kim YJ, Yang BS, Oh S, Han SJ, Lee HW, Heo J, et al. Photobias instability of high performance solution processed amorphous zinc tin oxide transistors. ACS applied materials & interfaces. 2013;5(8):3255-3261.

[28] Wang L-G, Qian X, Cao Y-Q, Cao Z-Y, Fang G-Y, Li A-D, et al. Excellent resistive switching properties of atomic layer-deposited Al_2O_3/HfO_2/Al_2O_3 trilayer structures for non-volatile memory applications. Nanoscale research letters. 2015;10(1):1-8.

[29] Kim Y-H, Heo J-S, Kim T-H, Park S, Yoon M-H, Kim J, et al. Flexible metal-oxide devices made by room-temperature photochemical activation of sol–gel films. Nature. 2012;489 (7414):128-132.

[30] Lian W, Long S, Lü H, Liu Q, Li Y, Zhang S, et al. Approaches for improving the performance of filament-type resistive switching memory. Chinese Science Bulletin. 2011;56(4): 461-464.

[31] Long S, Lian X, Cagli C, Cartoixa X, Rurali R, Miranda E, et al. Quantum-size effects in hafnium-oxide resistive

switching. Applied Physics Letters. 2013;102(18):183505.

[32] Kumar S, Wang Z, Huang X, Kumari N, Davila N, Strachan JP, et al. Oxygen migration during resistance switching and failure of hafnium oxide memristors. Applied Physics Letters. 2017;110(10):103503.

[33] Chen L, Dai Y-W, Sun Q-Q, Guo J-J, Zhou P, Zhang DW. Al_2O_3/HfO_2 functional stack films based resistive switching memories with controlled SET and RESET voltages. Solid State Ionics. 2015;273:66-69.

[34] Long S, Perniola L, Cagli C, Buckley J, Lian X, Miranda E, et al. Voltage and power-controlled regimes in the progressive unipolar RESET transition of HfO_2-based RRAM. Scientific reports. 2013;3(1):1-8.

[35] Sungjun, Ki, Hyungjin, Kim, Sungmin, Hwang, et al. Analog Synaptic Behavior of a Silicon Nitride Memristor. ACS Applied Materials & Interfaces. 2017;9(46).

[36] Yu S, Wu Y, Jeyasingh R, Kuzum D, Wong H-SP. An electronic synapse device based on metal oxide resistive switching memory for neuromorphic computation. IEEE Transactions on Electron Devices. 2011;58(8):2729-2737.

[37] Huang XD, Li Y, Li HY, Xue KH, Miao XS. Forming-Free, Fast, Uniform, and High Endurance Resistive Switching From Cryogenic to High Temperatures in $W/AlO_x/Al_2O_3/Pt$ Bilayer Memristor. IEEE Electron Device Letters. 2020;PP(99):1-.

[38] Yan X, Qin C, Lu C, Zhao J, Li H. Robust $Ag/ZrO_2/WS_2/Pt$ Memristor for Neuromorphic Computing. ACS Applied Materials & Interfaces. 2019;11: 48029-48038.

Section 2

Memristor - Advance Applications

Chapter 4

Mitigating State-Drift in Memristor Crossbar Arrays for Vector Matrix Multiplication

Amirali Amirsoleimani, Tony Liu, Fabien Alibart, Serge Eccofey, Yao-Feng Chang, Dominique Drouin and Roman Genov

Abstract

In this Chapter, we review the recent progress on resistance drift mitigation techniques for resistive switching memory devices (specifically memristors) and its impact on the accuracy in deep neural network applications. In the first section of the chapter, we investigate the importance of soft errors and their detrimental impact on memristor-based vector–matrix multiplication (VMM) platforms performance specially the memristance state-drift induced by long-term recurring inference operations with sub-threshold stress voltage. Also, we briefly review some currently developed state-drift mitigation methods. In the next section of the chapter, we will discuss an adaptive inference technique with low hardware overhead to mitigate the memristance drift in memristive VMM platform by using optimization techniques to adjust the inference voltage characteristic associated with different network layers. Also, we present simulation results and performance improvements achieved by applying the proposed inference technique by considering non-idealities for various deep network applications on memristor crossbar arrays. This chapter suggests that a simple low overhead inference technique can revive the functionality, enhance the performance of memristor-based VMM arrays and significantly increases their lifetime which can be a very important factor toward making this technology as a main stream player in future in-memory computing platforms.

Keywords: Memristor Crossbar, State-drift, Vector–matrix Multiplication, Inference

1. Introduction

Designing specialized hardware accelerators has been a topic of interest recently due to rapid growth of machine learning and artificial intelligence [1–5]. Despite the recent advancements in developing machine learning complementary-metal-oxide (CMOS) based chips to efficiently implement vector–matrix multiplication (VMM) operations, these systems are limited by the off-chip memory bottleneck [6]. To this end, co-locating memory and processing has been considered as a solution by using non-volatile resistive switching memory technologies [7–14]. One such device technology is memristor and it has risen as a promising high speed, low power

 IntechOpen

computational alternative to traditional CMOS hardware [15–18]. Memristor can be placed in a crossbar structure to perform highly parallel multiply-accumulate (MAC) operations efficiently using Ohm's Law [19–22]. Through heavy parallelization using Kirchoff's laws, a memristor crossbar is able to do MAC operations at $O(1)$ speed [23]. Crossbars are most commonly used to perform VMM by mapping a n by m matrix the memristors' conductance range, applying an input voltage vector to the rows, and then reading the current from the appropriate columns [24]. In addition to neuromorphic applications [25–28] of memristor crossbar, through VMM memristor crossbar has shown to be capable of performing a number of tasks ranging from image processing [29], physical unclonable functions (PUFs) [30–32], optimization problems [33–35], sparse coding [36], and solving partial differential equations [37]. Also, there have been many researches focusing on implementation of deep neural network (DNN) accelerators using memristor crossbars which focuses on different device, algorithm and system level contributions [18–20, 38]. While there has been significant progress in memristor crossbar-based computational devices, there are still many major challenges in robustness and computational accuracy that hinder the technology [20, 39, 40]. There have been several researches focus on mitigating the impact of non-idealities on memristor crossbar systems performance and these reliability improvement techniques are mainly proposed for process variations [41, 42], hard faults [43, 44], signal distortion issue [45], and memristance drift [46–48]. These techniques can be mainly categorized into (i) retraining to compensate the error (ii) mapping techniques (iii) closed-loop training (iv) error-correction coding.

Here, we focus on the memristance drift effect and the mitigation techniques to avoid the impact of this issue over the memrisitive DNN systems. Typically, in memristor crossbars, there are two major operations to perform: write and read operations. In the write operation, a voltage above the switching voltage of the memristor [49] is applied to a memristor repeatedly until the resistance of the memristor is sufficiently close to the target resistance. During the read operation, a voltage lower than the switching voltage is applied to the memristor and the current from the memristor is measured. The read operation is used extensively in the inference operation of many Ex-situ and In-situ algorithms including artificial neural networks (ANN). Ideally, the resistance of the memristor should not change at all, but in practice, there is often a very small change in the memristor state after a read operation. This phenomenon is known as memristance drift [50, 51]. Over many read operations, these small changes in resistance of the memristors in a crossbar will add up to have a significant impact on computational accuracy. Memristance drift occurs in different resistive switching memory technologies and it is not similar in terms of the behavior. In phase change memory (PCM) devices, the memristance drift occurs even when there is no voltage applied over the memory cell and the amorphous state (high resistance state) of the device is changing over time [52]. Subsequently, this issue will be more severe a the high-resistive amorphous state increases and this will impact dramatically the PCM-based system's performance in presence of high cycle-to-cycle and device-to-device variations. In memristor technology, as discussed before, the repetitive VMM operations result a memristance drift phenomenon and it becomes worse as the number of inference operations increases. Existing solutions to this problem include periodical weight reprogramming and feedback designs have limitations with high computational overhead and limited long-term effectiveness. For instance, HfOx RRAM testing results using a dynamic BL-bias circuit for preventing memristor state disturbance during read operations [53]. Error correction code (ECC) is used in [51] to reduce write latency by up to 70%. However, these techniques are not sufficient in themselves to enhances the performance of computational memristor crossbars in which 2–3 bits of memristance drift can cause significant decreases in performance. Recently, a few more effective memristance-drift mitigation techniques have been proposed [54–56]. This chapter will summarize the approach and results of a closed-loop feedback system technique [54] and an inline calibration approach [55] before taking a more in-depth look into an adaptive inference technique (AIDX) [56] that optimizes the inference voltage pulse amplitude and width. In addition, the power and chip area overhead of these three techniques are briefly compared.

2. Memristance drift and its modeling

In general, there are memristor models can be separated into physics-based and behavior-based simulations depending on the characteristics of their modeling and their general purpose. Physics-based models typically attempt to simulate memristors at a molecular-level by considering the material characteristics of the active memristor layers and mathematical modeling of the ion drifting between these materials. While physical models accurately model memristance drift, they are generally computationally expensive and limited in scope to the detailed analysis of singular memristor behavior. As such, memristor crossbar arrays are modeled using behavior-based models. Behavior-based memristor models are much simpler than physics-based models and use experimental fitting parameters to match the behavior of different types of memristors. Current–voltage plots are one of the most common methods of quickly visualizing memristor short-term behavior and many behavior-based models like VTEAM [57] are built to agree with these plots. Over the course of a voltage sweep, there is not enough time for the memristor's state to change noticeably under the threshold voltage and as such, the long term consequences of memristance drift aren't captured in these models. Many popular behavior-based models, such as VTEAM [57] and TEAM [58], utilize current and voltage thresholds to partition memristor behavior under high and low voltage/current scenarios. Generally, these threshold models approximate the subthreshold state change as zero and thus do not consider the long-term effects of memristance drift. Other behavior-models, such as the nonlinear ion drift [59] and Simmons Tunnel Barrier model [60], do not utilize a threshold and instead model memristor high and low voltage memristor behavior using the same equations and fitting parameters. Without a threshold, these models lack the flexibility to accurately model the minute changes of memristance drift without sacrificing the accuracy of its higher-voltage switching modeling. Recently, there have been attempts to extend popular behavior-based models to more accurately simulate memristance drift. For instance, [56] added subthreshold modeling equation and fitting parameters to extend the VTEAM model. However, the modeling of memristance drift in behavior-based models is still currently in its infancy due to the lack of experimental data on long-term memristor behavior when exposed to low voltage pulses.

2.1 Impact of state-drift on crossbar VMM

Memristance drift in crossbar arrays can be summarized as the buildup of small unintended changes in memristor state over many low-voltage read operations [56]. During a crossbar VMM operation, the ideal output current I_j of the j-th column can be modeled simply as:

$$I_j = \sum_i G_{ij} V_i \tag{1}$$

Here V_i represents the voltage applied to the i-th row of the crossbar and G_{ij} is the conductance of the memristors in the i-th row and j-th column. For a given VMM operation, state drift can be represented as a small change in conductance δG. The non-ideal output current I'_j is then given as:

$$I'_j = \sum_i (G_{ij} + \Delta G_{ij}) V_i \tag{2}$$

ΔG_{ij} represents the accumulation of memristance drift caused by all previous VMM operations and skews the distribution of weights within the crossbar (**Figure 1a**). It can be shown from Eqs. (1) and (2) that the difference between I'_j and I_j scales with the read voltage of each row V and the number of rows N. Over a long period, the buildup of drift in the crossbar weights caused by memristance drift will lead to significant error (**Figure 1b**). The speed of memristance drift will vary depending on application and memristor characteristics. Even after a short period of 1 second, a 0.1 V signal could cause a memristor to deviate around 2% from its initial state [61]. A thorough analysis of memristance drift speed with respect to initial state and drift direction is done in [54]. In the SET direction, [54] found that memristors' resistance decreased by 77.07%, 62.07%, 56.28% and 8.81% after 100 read operations with initial resistances of 200kΩ, 100kΩ, 80kΩ, and 15kΩ, respectively. The speed of drift in the RESET direction was much slower at only ~0%, 1.17 × 10–4%, 0.018% and 16.43% increase in resistance for the same initial states. From the analysis in [54], memristance drift speed is shown to be greatly impacted by initial state and the direction of switching. A heatmap of conductance highlights the impact of initial state on the buildup of memristance drift over time

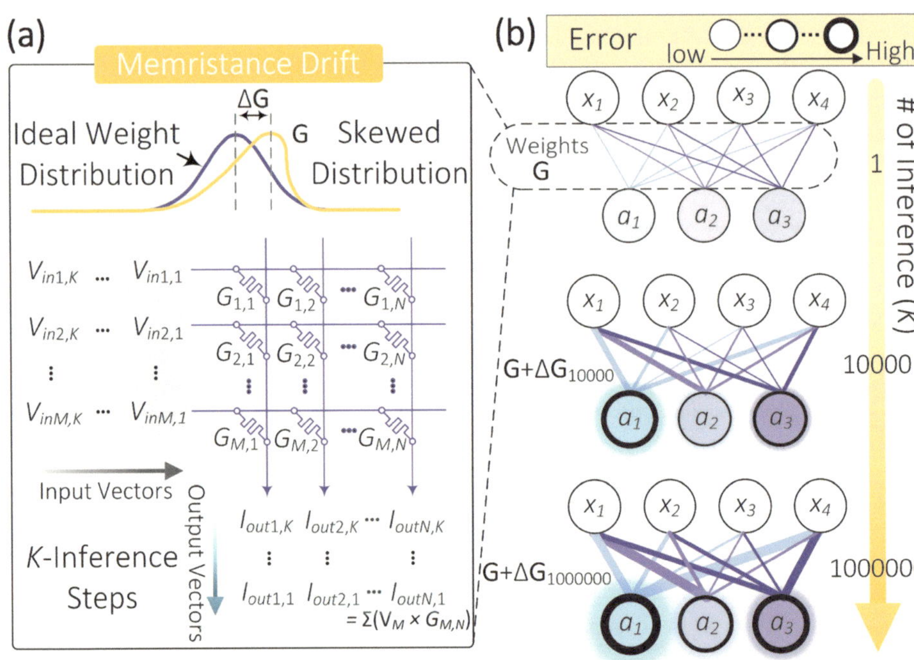

Figure 1.
(a) Neural network (NN) weight matrix mapped onto crossbar with conductance matrix G. Each subsequent VMM operation will cause a slight skew in the distribution of the memristors' conductance. (b) Initially, the error due to memristance drift is negligible. Over many inference operations, the skew in the conductance distribution builds up resulting in significant error in the NN output. Figure reprinted by [56].

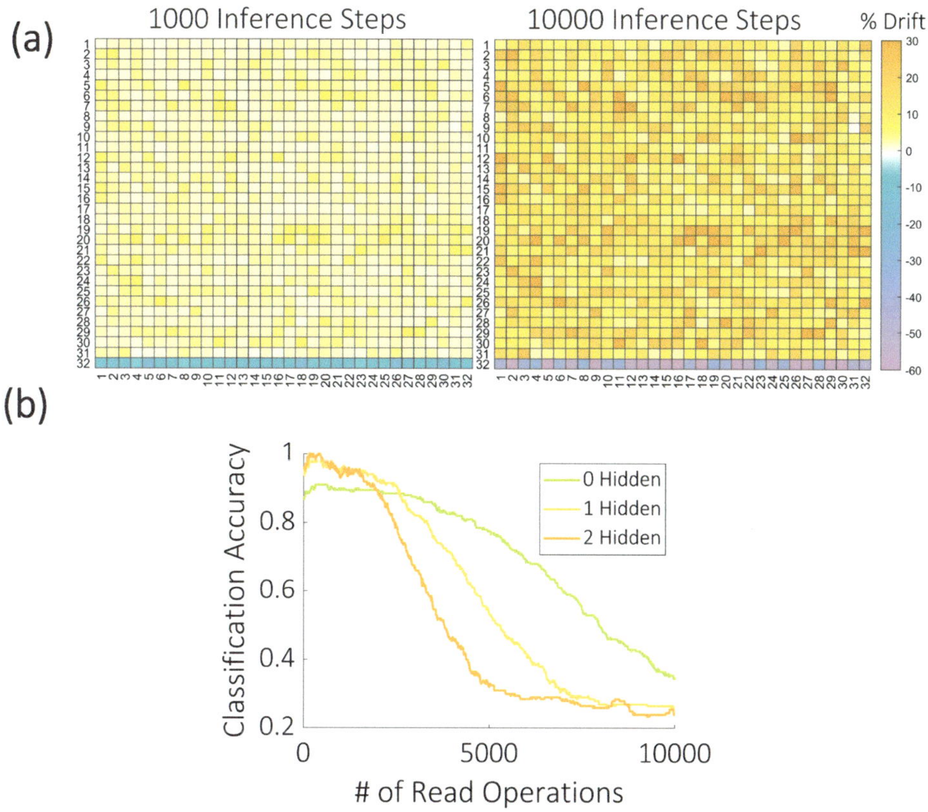

Figure 2.
(a) Impact of varying number of hidden layers on memristance drift induced accuracy degradation on the MNIST dataset [62]. (b) Heatmap of typical memristor crossbar weight mapping with the bottom row as the bias and the rest of the heatmap representing the weight matrix. Over time, the memristance drift direction between the bias and weights diverges due to differences in bias and weight initialization. Figure reprinted by [56].

(**Figure 2a**). Here, the bottom row represents the bias of a neural network and are initialized near the high conductance state while the rest of the memristors representing the weights are initialized close to the low conductance state. Due to this conductance initialization, all memristors except the bottom row in the above figure experience an aggregate memristance drift in the positive direction while the bottom row experiences memristance drift in the negative direction. There have been multiple studies that show that memristance drift causes significant performance degradation on various applications after long term use. In [56], the negative impact across ten baseline ML tasks in the Proben1 [63] datasets, memristance drift caused an average classification accuracy decrease of 26%, 42%, and 51% after 500, 2000, and 10000 inference operations, respectively. [56] also analyzed the effects of memristance drift on convolutional neural networks with the CIFAR10 image classification dataset [64]. Ten different CNN architectures were tested with a relatively consistent accuracy degradation of 29%, 59%, and 72% after 500, 2000, and 10000 inference operations respectively. Memristor drift ranges from upwards of 10% deviatation from its programmed value to upwards of 30% deviation at 10000 inference steps as shown in (**Figure 2a**). To clarify, the memristance drift speed remains the same for each network in (**Figure 2b**) regardless of the number of hidden layers. However, each additional hidden layer accumulates memristance increasing amounts of memristance drift-related error from the previous layer causing deeper neural network's accuracy to degrade more quickly than networks with less hidden layers (**Figure 2b**). In [55], the classification accuracy on the

MNIST handwritten digits dataset [62] decreases by approximately 2.5–4% across four independent trials due to the cycle-to-cycle variation of memristance drift. In [54], the classification accuracy degradation on MNIST was tested with memristance drift in the SET and RESET directions separately. In the SET direction, the classification accuracy dropped from 91.91–60% after only 100 inference operations. In the RESET direction, accuracy degradation was slower where accuracy dropped to 60% after approximately 300 inference operations.

3. Memristance drift mitigation overview

3.1 ICE: inline calibration

Given the significant negative impact of memristance drift on crossbar performance, there has been a few works that have proposed meaningful solutions to the memristance drift problem. When a memristor crossbar's performance drops below acceptable levels due to memristance drift, the memristors will be rewritten to their intended states. For a given application, this recalibration is usually done at periodically ensure high crossbar accuracy over long time intervals. Since rewriting a memristor to a specific state can be up to 100 times slower than the speed of an inference operation [65], frequent crossbar calibrations could significantly bottleneck crossbar throughput. In [55], the authors propose an inline calibration method that utilizes "interrupt-and-benchmark (I&B)" operations to track the crossbar computational error in order to predict the time period before the next calibration operation. In addition, the time period between two I&B operations is dynamically optimized as to minimize the time overhead of the inline calibration method on crossbar performance. As defined in the paper, I&B operations interrupt regular crossbar operation in order to evaluate crossbar computational error on a set of benchmark data [55]. Between each recalibration of the memristor crossbar, there exists a theoretical maximum number of inference operations $sup\ n_r$ before the next recalibration must be done due to performance degradation. The optimization goal of ICE is to make the actual number of inference operations n_r between two calibration operations approach $sup\ n_r$. This is not a trivial task because $sup\ n_r$ can vary significantly between many calibration operations due to changes in memristance drift speed from cycle-to-cycle variations and other factors. ICE approximates $sup\ n_r$ by applying polynomial fitting to the crossbar error data during I&B operations.

To reduce the time overhead of the inline calibration method, ICE seeks to minimize the number of I&B operations k between each recalibration of the crossbar. ICE uses its polynomial fitting function to guess the benchmark computation error of the next I&B operation. If the absolute difference between the guessed crossbar error and the actual I&B error is below some threshold, the time until the next I&B operation will be doubled up to some maximum time interval. Otherwise, ICE will reset the time interval between successive I&B operations to its default value. For testing, ICE adopts a TiOx-based memristor device model from [61]. This model extends the bulk model TiO2 with a focus on behavior-based process variation analysis of memristors. The results presented in [55] are measured in terms of efficiency with computational efficiency defined as:

$$\gamma = \frac{n_r}{sup\ n_r} \tag{3}$$

To quantify the time overhead caused by I&B operations, the parameter Δ is defined as:

$$\Delta = \frac{kn_{IB}}{n_r} \tag{4}$$

Here, k is the average number of I&B operations between crossbar recalibrations and n_{IB} is the number of inference operations required per I&B operation. These criteria were evaluated across four baseline tasks (HMAX, KMeans, Sobel, and MNIST [62]) and compared to a baseline of constant time period crossbar recalibrations. With second degree polynomial fitting, ICE achieves an average calibration efficiency γ of 91.18% which is a 21.77% improvement over the baseline. This improvement in calibration efficiency only came at the cost of a time overhead Δ of 0.439%. No information on the voltage range or power consumption of ICE is presented in [55]. However, the author's mentioned that future works could include a detailed power analysis of ICE using a SPICE-based memristor model.

3.2 Closed-loop feedback circuit

Traditionally, the data flow of a memristor crossbar-based neural network follows a linear pipeline in a conventional open-loop system. While these open-loop systems serve as a simple and efficient pipeline for VMM operations, they are not able to effectively manage the effects of memristance drift. In [54], a closed-loop circuit is proposed to mitigate memristance drift by adaptively adjusting the direction of current in each memristor using a feedback controller. Mean square error (MSE) is used to measure the degradation caused by memristance drift. Specifically, the difference in MSE (ΔMSE) between the ideal crossbar and the current state is used as a metric to inform the feedback controller. For each inference operation, a weight compensation algorithm is run to minimize memristance drift speed. The second feature of the closed-loop design introduced in [54] is the usage of an "arrogant principle" which assumes that the prediction made by the crossbar system is always correct. This principle allows the system to use its output as the label to determine the direction of compensation in the weight compensation algorithm. The effectiveness of this assumption hinges on the ideal accuracy of the crossbar-mapped neural network. With a high initial accuracy, the "expectation of recognition accuracy probability with respect to time" will be close to its upper bound, thus keeping the rate of degradation of the feedback controller low. Naively, this closed-loop design requires an additional compensation pulse for each inference operation which would halve the throughput of the crossbar system. To address this issue, [54] combines the compensation pulse with regular inference operations by manipulating the k-th inference operation into compensating for the memristance drift caused by the (k-1)-th inference. The feedback controller is used to determine the recall direction of the (k-1)-th and then adjusts the direction of the k-th inference pulse accordingly. By integrating the k-th compensation pulse into the (k + 1)-th inference pulse, there is no need to sacrifice crossbar throughput to implement this proposed system. For testing, [54] adopts the dynamic model of the TaOx memristor from [66] with a low resistive state and high resistive state of 1 kΩ and 1MΩ, respectively. The performance of single and two-layer neural networks were tested on the MNIST dataset. The effectiveness of the closed loop design was measured as the number of inference operations before the crossbar accuracy drops below 70%. As compared to a baseline crossbar system, the proposed closed-loop design is shown to increase the number of recall operations by 1897 operations for the single-layer network and 1590 operations for the two-layer network. This increase corresponds to a 13.84× lifetime extension for the single layer network and a 13.95× lifetime extension for the two-layer network. For power estimation, [54] uses a square recall voltage pulse of 0.3 V for 100 ns. Boban had an estimated power consumption of 1.1196 mW for a single layer neural network which increases to 6.7367 mW for a two-layer neural network implementation.

4. Adaptive inference scheme for memristance drift mitigation

In [56], the authors proposed an adaptive inference scheme called AIDX that optimizes the amplitude and duration of inference voltage pulses in order to minimize the speed of memristance drift. AIDX formulates memristance drift as an optimization problem and seeks to minimize the memristance drift error defined as the increase in mean squared error (MSE) from the initial programmed crossbar after a set number of inference operations. The initial MSE can be modeled as:

$$E_0 = \sum_j \left(y_j - \sum_i G_{ij} V_i \right)^2 \tag{5}$$

Similarly, the MSE after k inference operations is given as:

$$E_k = \sum_j \left(y_j - \sum_i \left(G_{ij} + \Delta G_{ij}^k \right) V_i \right)^2 \tag{6}$$

where ΔG_{ij}^k is the accumulated memristance drift in the memristor i-th row and j-th column from k inference operations. The additional error due to memristance drift after k operations is calculated as $E_{Drift} = E_k - E_0$. The naive approach of simply choosing the minimum allowable voltage amplitude and duration may seem logical because speed of memristance drift scales with amplitude and duration. However, this naïve approach would still result in significant memristance drift because of the vast differences in state drift speed in the SET and RESET drift [54]. As such, AIDX focuses on balancing the aggregate drift in the SET and RESET directions for a given application by optimizing the ratio of SET to RESET voltage pulse amplitude $\boldsymbol{A}$ and duration $\boldsymbol{D}$ (**Figure 3a**). The minimization of memristance drift with respect to voltage pulse amplitude and duration is formalized as follows:

$$\mathrm{min}_{\mathrm{A,D}} \mathrm{E}_{\mathrm{Drift}}(\mathbf{A}, \mathbf{D}) \tag{7}$$

Here, E_{Drift} is a function of **A** and **D** that are vectors which represent the ratio of SET to RESET voltage pulse amplitude and duration of each row of the memristor crossbar respectively. The Broyden-Fletcher-Goldfarb-Shannon (BFGS) algorithm [67] is used to tackle this optimization problem. Since the gradient of the memristance drift error cannot be evaluated directly, the gradients used in the BFGS algorithm were numerically approximated using function evaluations. Under some circumstances, it is possible for the optimized values of **A** and **D** to be too large or small to be properly implemented on a crossbar. Extreme values of **A** and **D** are often caused by skewed memristor characteristics where memristance drift speed in one direction is much faster than the other direction or when the data distribution is heavily skewed toward one recall direction. To address this issue, AIDX randomly inverting the direction of inference for an input vector **x** with probability a in order to compensate for imbalanced memristor characteristics and a skewed data distribution (**Figure 3b**). The probability of input inversion is optimized to minimize E_{Drift} before the optimization of **A** and **D** vectors to ensure a

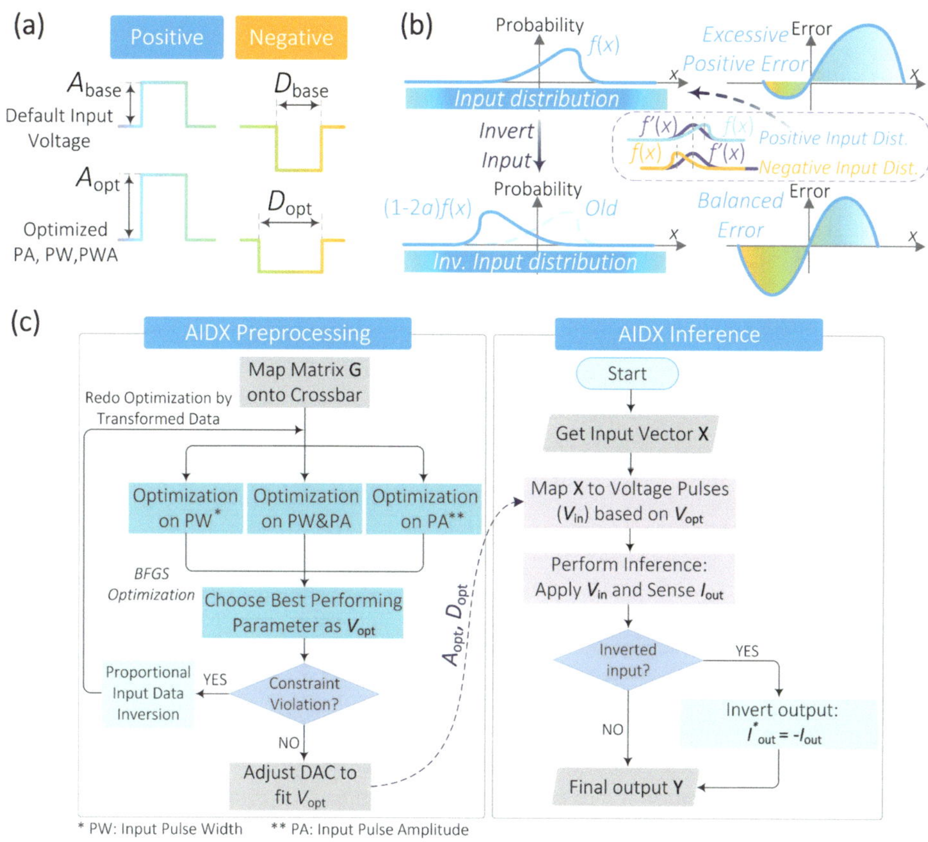

Figure 3.
(a) Visual representation on how AIDX's parameters changes the relative amplitude and width of positive and negative voltage pulses before and after optimization. (b) Illustration of skewed input distributions can cause an imbalance of memristance drift error in a particular direction. By applying input inversion, the input distribution is reflected such that the memristance drift error in each direction is balanced. (c) Flowchart describing the AIDX procedure for preprocessing and inference. Figure reprinted by [56].

relatively balanced memristance drift speed in the SET and RESET directions as to prevent extreme final values of **A** and **D.**

The general usage of AIDX in both preprocessing and inference is described in (**Figure 3c**). To ensure optimal performance, optimization is done in three scenarios: Optimizing over pulse amplitude ratio, optimizing over pulse duration ration, and optimization over both parameters simultaneously. The best set of parameters is then chosen according to lowest evaluated E_{Drift}. If the parameters **A** and **D** are too extreme, the optimization of probability of input inversion a is performed. Applying AIDX during inference is almost identical to a normal crossbar VMM operation except that if the input was inverted, the output vector must be reinverted to recover the intended output. When applying AIDX to deep neural networks, it can be inefficient to optimize over all layers simultaneously due to the large number of parameters. Instead, AIDX applies the BFGS algorithm to each layer separately in forward pass order in order to reduce optimization time (**Figure 4a**). AIDX used a simulated extended VTEAM model to fit real TiOx-based memristor device data. The following non-idealities were considered for testing: 15% memristor programming error, 15% random gaussian noise added to the high/low conductance states and alpha/k parameters of the extended VTEAM model for device-to-device variation. In addition, 200 ohms of source resistance and 20 ohms of line resistance were considered as well as sneak paths were also considered. AIDX was applied to a memristor system that

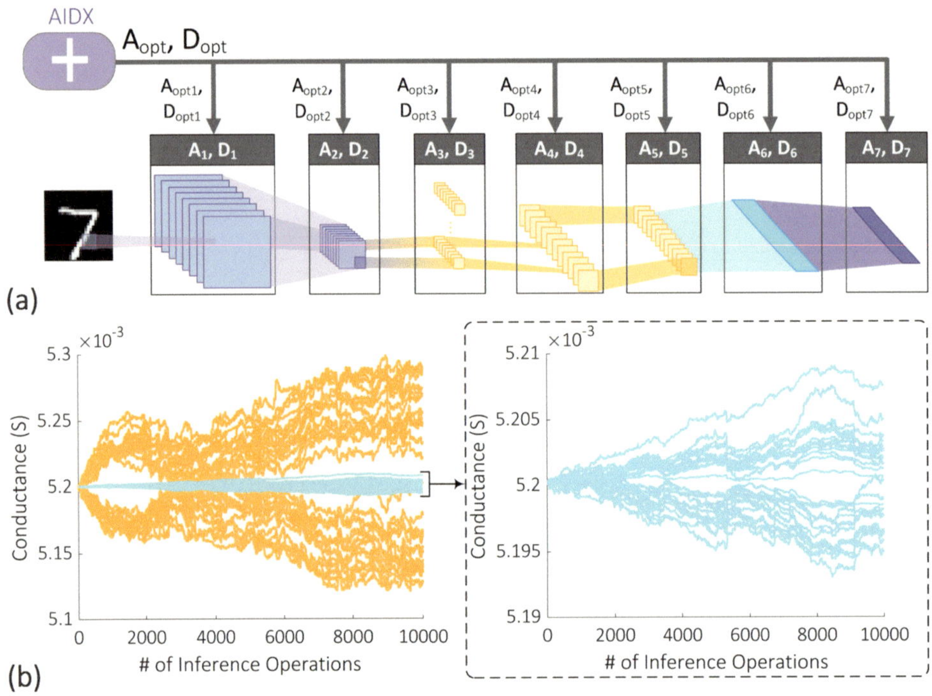

Figure 4.
(a) AIDX optimizes and applies separate voltage amplitude and duration ratios for each layer separately. (b) Basic test on memristance drift with all devices with and without AIDX. With all memristors initially set to 0.0052 S, half of the memristors receive pre-generated random sequences of positive and an identical sequence of negative pulses are applied to the other half of the crossbar.

operates with a voltage range of −0.2 V to 0.2 V. Depending on the pulse amplitude optimization, AIDX's inference pulse can vary within this range. On average, there is 2% reduction in passive power consumption within the crossbar compared to the baseline system when using AIDX. A simple test of AIDX effectiveness is done by applying random positive and negative voltage pulses to memristors with and without AIDX. After 10000 inference operations, the baseline memristors deviated a max of 1.9% from its initial value while AIDX had a max deviation of only 0.17% (**Figure 4b**). When tested on the 10 benchmark tasks from the Proben1 dataset [63], the average classification accuracy degradation with AIDX is approximately 4%, 7%,

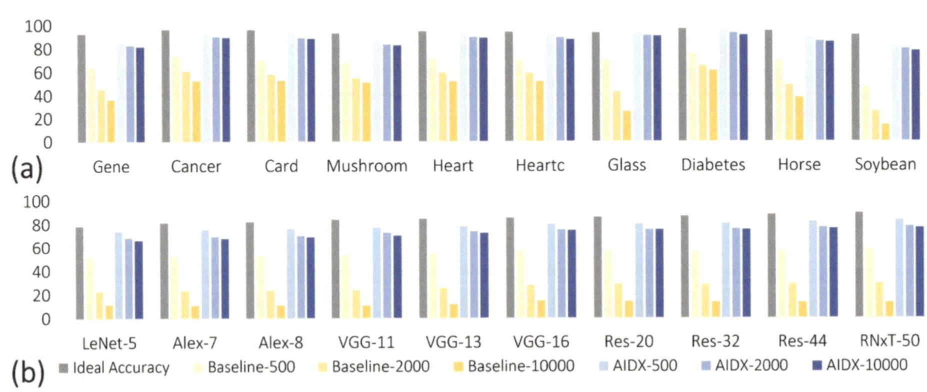

Figure 5.
(a) Classification accuracy of AIDX and baseline neural network on ten tasks from the Proben1 dataset after 500, 2000, and 10000 inference operations. (b) Classification accuracy of AIDX and baseline on CIFAR10 dataset with various CNN architectures after 500, 2000, and 10000 inference operations. Figure reprinted by [56].

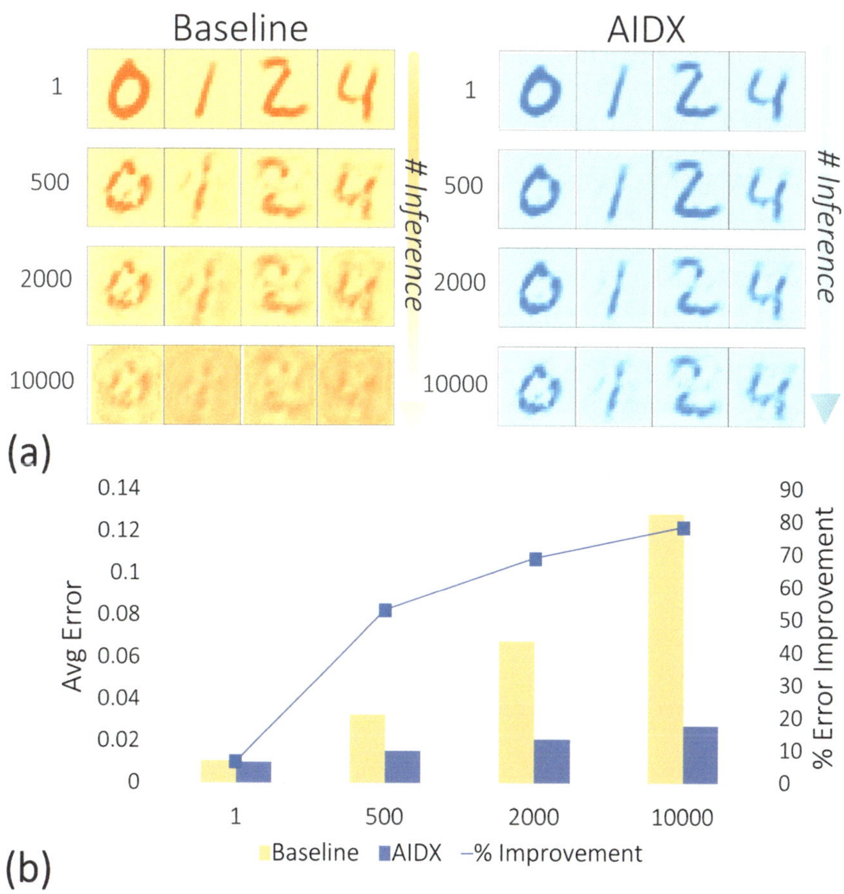

Figure 6.
(a) Reconstruction of sample images from MNIST dataset after 1, 500, 2000, and 10000 inference operations. (b) The average mean squared error in image reconstruction between AIDX and baseline autoencoder after set time steps. The percentage error improvement of AIDX over the baseline is also shown. Figure reprinted by [56].

and 8% after 500, 2000, and 10000 inference operations (**Figure 5a**). On average, AIDX reduced accuracy degradation by 42% as compared to the baseline test after 10000 inference operations. When testing AIDX on 10 CNN architectures using the CIFAR10 dataset [18], the classification accuracy decrease by an average of 4%, 7%, and 8% after 500, 2000, and 10000 inference operations (**Figure 5b**). This accuracy degradation corresponds to a 22%, 35%, and 43% improvement over the baseline respectively. In addition, AIDX was also applied to image reconstruction by training a simple 3-layer autoencoder on the MNIST dataset [62]. The average mean squared error of the baseline auto-encoder was 0.033, 0.068, and 0.129 after 500, 2000, and 10000 inference operations respectively. With AIDX, the average mean squared error drops to 0.015, 0.021, and 0.028 after 500, 2000, and 10000 inference operations which is an improvement of 53.0%, 69.0%, and 78.6% over the baseline (**Figure 6**).

5. Overhead analysis

The three methods for mitigating memristance drift discussed in this chapter all induce small overheads in terms of power consumption and chip area. Time overhead is not discussed in this section because there is negligible change in crossbar throughput by all three mitigation methods. Power overhead is defined in this section as the additional power consumption induced in the memristor crossbar and peripheral circuits due to proposed memristance drift solutions. For the sake of consistency, the estimates of peripheral power consumption of [54] are used for comparison. While power consumption is not disclosed in [55], the power overhead of [56] is 1.19% while [54] has a power overhead of 1.61%. Area overhead is defined consistently with [56] as the additional on-chip area required for memristance drift mitigation method because of peripherals, external circuit, and other items. Since both [55, 56] do not include any additional on-chip circuitry, these two methods do not have any chip area overhead while the closed loop circuits proposed in [54] require an additional 2.34% chip area. On the other hand, both [55, 56] require solving an optimization problem before implementing their mitigation technique. However, considering that the optimization procedure would only needed to be performed once for an application, these solutions still promise great scalability for long-term memristor crossbar usage.

6. Conclusions

In summary, this chapter first discusses memristor crossbar modeling and how there is a current lack of attention in modeling subthreshold memristor behavior. The next section overviews how the speed of memristance drift is impacted by recall voltage and amplitude, memristor characteristics, crossbar size, and number of inference operations since the last write operation. In addition, memristance drift is shown to cause severe accuracy degradation across multiple datasets and tasks such as MNIST and CIFAR10. The second half of this chapter is dedicated to overviewing three different approaches for memristance drift mitigation. First, an inline calibration approach [55] and a closed-loop feedback system is summarized. Then, there is a more in-depth look into an adaptive inference scheme that optimized the ratio of SET to RESET voltage pulse amplitude and width to minimize memristance drift speed. The final section of the chapter briefly compared the power and chip area overhead of these three memristance drift mitigation techniques. Hopefully, this chapter can bring more much-needed attention to the study of memristance drift and the development of drift mitigation techniques.

Author details

Amirali Amirsoleimani[1*], Tony Liu[2], Fabien Alibart[3], Serge Eccofey[3], Yao-Feng Chang[4], Dominique Drouin[3] and Roman Genov[2]

1 Department of Electrical Engineering and Computer Science, Lassonde School of Engineering, York University, Toronto, Ontario, Canada

2 Intelligent Sensory Microsystem Laboratory, University of Toronto, Toronto, Ontario, Canada

3 Interdisciplinary Institute for Technological Innovation – 3IT, University of Sherbrooke, Sherbrooke, Quebec, Canada

4 Microelectronics Research Center, The University of Texas at Austin, Austin, TX, USA

*Address all correspondence to: amirsol@yorku.ca

References

[1] Jouppi NP, Young C, Patil N, Patterson D, Agrawal G, Bajwa R, Bates S, Bhatia S, Boden N, Borchers A, Boyle R. In-datacenter performance analysis of a tensor processing unit. In Proceedings of the 44th annual international symposium on computer architecture 2017 Jun 24 (pp. 1-12).

[2] Chung E, Fowers J, Ovtcharov K, Papamichael M, Caulfield A, Massengill T, Liu M, Lo D, Alkalay S, Haselman M, Abeydeera M. Serving dnns in real time at datacenter scale with project brainwave. IEEE Micro. 2018 Apr 20;38(2):8-20.

[3] Chen YH, Krishna T, Emer JS, Sze V. Eyeriss: An energy-efficient reconfigurable accelerator for deep convolutional neural networks. IEEE journal of solid-state circuits. 2016 Nov 8;52(1):127-138.

[4] Lee J, Kim C, Kang S, Shin D, Kim S, Yoo HJ. UNPU: A 50.6 TOPS/W unified deep neural network accelerator with 1b-to-16b fully-variable weight bit-precision. In2018 IEEE International Solid-State Circuits Conference-(ISSCC) 2018 Feb 11 (pp. 218-220). IEEE.

[5] Moons B, Uytterhoeven R, Dehaene W, Verhelst M. 14.5 envision: A 0.26-to-10tops/w subword-parallel dynamic-voltage-accuracy-frequency-scalable convolutional neural network processor in 28nm fdsoi. In2017 IEEE International Solid-State Circuits Conference (ISSCC) 2017 Feb 5 (pp. 246-247). IEEE.

[6] Wulf WA, McKee SA. Hitting the memory wall: Implications of the obvious. ACM SIGARCH computer architecture news. 1995 Mar 1;23(1): 20-24.

[7] Yu S. Neuro-inspired computing with emerging nonvolatile memorys. Proceedings of the IEEE. 2018 Feb;106 (2):260-285.

[8] Chakraborty I, Jaiswal A, Saha AK, Gupta SK, Roy K. Pathways to efficient neuromorphic computing with non-volatile memory technologies. Applied Physics Reviews. 2020 Jun 3;7(2):021308.

[9] Ambrogio S, Narayanan P, Tsai H, Shelby RM, Boybat I, Di Nolfo C, Sidler S, Giordano M, Bodini M, Farinha NC, Killeen B. Equivalent-accuracy accelerated neural-network training using analogue memory. Nature. 2018 Jun;558(7708):60-67.

[10] Hu M, Graves CE, Li C, Li Y, Ge N, Montgomery E, Davila N, Jiang H, Williams RS, Yang JJ, Xia Q. Memristor-based analog computation and neural network classification with a dot product engine. Advanced Materials. 2018 Mar;30(9):1705914.

[11] Cai F, Correll JM, Lee SH, Lim Y, Bothra V, Zhang Z, Flynn MP, Lu WD. A fully integrated reprogrammable memristor–CMOS system for efficient multiply–accumulate operations. Nature Electronics. 2019 Jul;2(7):290-299.

[12] Li C, Belkin D, Li Y, Yan P, Hu M, Ge N, Jiang H, Montgomery E, Lin P, Wang Z, Song W. Efficient and self-adaptive in-situ learning in multilayer memristor neural networks. Nature communications. 2018 Jun 19;9(1):1-8.

[13] Ramasubramanian SG, Venkatesan R, Sharad M, Roy K, Raghunathan A. SPINDLE: SPINtronic deep learning engine for large-scale neuromorphic computing. In2014 IEEE/ ACM International Symposium on Low Power Electronics and Design (ISLPED) 2014 Aug 11 (pp. 15-20). IEEE.

[14] Ankit A, Hajj IE, Chalamalasetti SR, Ndu G, Foltin M, Williams RS, Faraboschi P, Hwu WM, Strachan JP,

Roy K, Milojicic DS. PUMA: A programmable ultra-efficient memristor-based accelerator for machine learning inference. In Proceedings of the Twenty-Fourth International Conference on Architectural Support for Programming Languages and Operating Systems 2019 Apr 4 (pp. 715-731).

[15] Hu M, Graves CE, Li C, Li Y, Ge N, Montgomery E, Davila N, Jiang H, Williams RS, Yang JJ, Xia Q. Memristor-based analog computation and neural network classification with a dot product engine. Advanced Materials. 2018 Mar;30(9):1705914.

[16] Ielmini D, Wong HS. In-memory computing with resistive switching devices. Nature Electronics. 2018 Jun;1 (6):333-343.

[17] Zidan MA, Strachan JP, Lu WD. The future of electronics based on memristive systems. Nature electronics. 2018 Jan;1(1):22-29.

[18] Sebastian A, Le Gallo M, Khaddam-Aljameh R, Eleftheriou E. Memory devices and applications for in-memory computing. Nature nanotechnology. 2020 Jul;15(7):529-544.

[19] Tsai H, Ambrogio S, Narayanan P, Shelby RM, Burr GW. Recent progress in analog memory-based accelerators for deep learning. Journal of Physics D: Applied Physics. 2018 Jun 21;51(28): 283001.

[20] Amirsoleimani A, Alibart F, Yon V, Xu J, Pazhouhandeh MR, Ecoffey S, Beilliard Y, Genov R, Drouin D. In-Memory Vector-Matrix Multiplication in Monolithic Complementary Metal–Oxide–Semiconductor-Memristor Integrated Circuits: Design Choices, Challenges, and Perspectives. Advanced Intelligent Systems. 2020 Nov;2(11): 2000115.

[21] Shafiee A, Nag A, Muralimanohar N, Balasubramonian R, Strachan JP, Hu M, Williams RS, Srikumar V. ISAAC: A convolutional neural network accelerator with in-situ analog arithmetic in crossbars. ACM SIGARCH Computer Architecture News. 2016 Jun 18;44(3):14-26.

[22] Chi P, Li S, Xu C, Zhang T, Zhao J, Liu Y, Wang Y, Xie Y. Prime: A novel processing-in-memory architecture for neural network computation in reram-based main memory. ACM SIGARCH Computer Architecture News. 2016 Jun 18;44(3):27-39.

[23] Yao P, Wu H, Gao B, Tang J, Zhang Q, Zhang W, Yang JJ, Qian H. Fully hardware-implemented memristor convolutional neural network. Nature. 2020 Jan;577(7792):641-646.

[24] Yakopcic C, Alom MZ, Taha TM. Extremely parallel memristor crossbar architecture for convolutional neural network implementation. In2017 International Joint Conference on Neural Networks (IJCNN) 2017 May 14 (pp. 1696-1703). IEEE.

[25] Rahimi Azghadi M, Chen YC, Eshraghian JK, Chen J, Lin CY, Amirsoleimani A, Mehonic A, Kenyon AJ, Fowler B, Lee JC, Chang YF. Complementary Metal-Oxide Semiconductor and Memristive Hardware for Neuromorphic Computing. Advanced Intelligent Systems. 2020 May;2(5):1900189.

[26] Jo SH, Chang T, Ebong I, Bhadviya BB, Mazumder P, Lu W. Nanoscale memristor device as synapse in neuromorphic systems. Nano letters. 2010 Apr 14;10(4): 1297-1301.

[27] Kim KH, Gaba S, Wheeler D, Cruz-Albrecht JM, Hussain T, Srinivasa N, Lu W. A functional hybrid memristor crossbar-array/CMOS system for data storage and neuromorphic applications. Nano letters. 2012 Jan 11;12 (1):389-395.

[28] Prezioso M, Merrikh-Bayat F, Hoskins BD, Adam GC, Likharev KK, Strukov DB. Training and operation of an integrated neuromorphic network based on metal-oxide memristors. Nature. 2015 May;521(7550):61-64.

[29] Li C, Hu M, Li Y, Jiang H, Ge N, Montgomery E, Zhang J, Song W, Dávila N, Graves CE, Li Z. Analogue signal and image processing with large memristor crossbars. Nature electronics. 2018 Jan;1(1):52-59.

[30] Gao L, Chen PY, Liu R, Yu S. Physical unclonable function exploiting sneak paths in resistive cross-point array. IEEE Transactions on Electron Devices. 2016 Jun 21;63(8):3109-3115.

[31] Nili H, Adam GC, Hoskins B, Prezioso M, Kim J, Mahmoodi MR, Bayat FM, Kavehei O, Strukov DB. Hardware-intrinsic security primitives enabled by analogue state and nonlinear conductance variations in integrated memristors. Nature Electronics. 2018 Mar;1(3):197-202.

[32] Jiang H, Li C, Zhang R, Yan P, Lin P, Li Y, Yang JJ, Holcomb D, Xia Q. A provable key destruction scheme based on memristive crossbar arrays. Nature Electronics. 2018 Oct;1(10):548-554.

[33] Cai F, Kumar S, Van Vaerenbergh T, Liu R, Li C, Yu S, Xia Q, Yang JJ, Beausoleil R, Lu W, Strachan JP. Harnessing intrinsic noise in memristor Hopfield neural networks for combinatorial optimization. arXiv preprint arXiv:1903.11194. 2019 Mar 26.

[34] Bojnordi MN, Ipek E. Memristive boltzmann machine: A hardware accelerator for combinatorial optimization and deep learning. In2016 IEEE International Symposium on High Performance Computer Architecture (HPCA) 2016 Mar 12 (pp. 1-13). IEEE.

[35] Liu S, Wang Y, Fardad M, Varshney PK. A memristor-based optimization framework for artificial intelligence applications. IEEE Circuits and Systems Magazine. 2018 Feb 9;18 (1):29-44.

[36] Sheridan PM, Cai F, Du C, Ma W, Zhang Z, Lu WD. Sparse coding with memristor networks. Nature nanotechnology. 2017 Aug;12(8):784.

[37] Zidan MA, Jeong Y, Lee J, Chen B, Huang S, Kushner MJ, Lu WD. A general memristor-based partial differential equation solver. Nature Electronics. 2018 Jul;1(7):411-420.

[38] Chen A, Datta S, Hu XS, Niemier MT, Rosing TŠ, Yang JJ. A survey on architecture advances enabled by emerging beyond-CMOS technologies. IEEE Design & Test. 2019 Feb 28;36(3):46-68.

[39] Jain S, Ankit A, Chakraborty I, Gokmen T, Rasch M, Haensch W, Roy K, Raghunathan A. Neural network accelerator design with resistive crossbars: Opportunities and challenges. IBM Journal of Research and Development. 2019 Oct 11;63(6):10-11.

[40] Mittal S. A survey of ReRAM-based architectures for processing-in-memory and neural networks. Machine learning and knowledge extraction. 2019 Mar;1 (1):75-114.

[41] Cheng M, Xia L, Zhu Z, Cai Y, Xie Y, Wang Y, Yang H. Time: A training-in-memory architecture for memristor-based deep neural networks. In2017 54th ACM/EDAC/IEEE Design Automation Conference (DAC) 2017 Jun 18 (pp. 1-6). IEEE.

[42] Chen L, Li J, Chen Y, Deng Q, Shen J, Liang X, Jiang L. Accelerator-friendly neural-network training: Learning variations and defects in RRAM crossbar. InDesign, Automation & Test in Europe Conference & Exhibition (DATE), 2017 2017 Mar 27 (pp. 19-24). IEEE.

[43] Xia L, Liu M, Ning X, Chakrabarty K, Wang Y. Fault-tolerant training with on-line fault detection for RRAM-based neural computing systems. In Proceedings of the 54th Annual Design Automation Conference 2017 2017 Jun 18 (pp. 1-6).

[44] Liu C, Hu M, Strachan JP, Li H. Rescuing memristor-based neuromorphic design with high defects. In2017 54th ACM/EDAC/IEEE Design Automation Conference (DAC) 2017 Jun 18 (pp. 1-6). IEEE.

[45] Liu X, Mao M, Liu B, Li H, Chen Y, Li B, Wang Y, Jiang H, Barnell M, Wu Q, Yang J. RENO: A high-efficient reconfigurable neuromorphic computing accelerator design. In Proceedings of the 52nd Annual Design Automation Conference 2015 Jun 7 (pp. 1-6).

[46] Li B, Shan Y, Hu M, Wang Y, Chen Y, Yang H. Memristor-based approximated computation. In: International Symposium on Low Power Electronics and Design (ISLPED); 2013; Beijing, China. p. 242-247. DOI: 10.1109/ISLPED.2013.6629302.

[47] Yan B, Yang J, Wu Q, Chen Y, Li H. A closed-loop design to enhance weight stability of memristor based neural network chips. In: 2017 IEEE/ACM International Conference on Computer-Aided Design (ICCAD); 2017; Irvine, CA, USA. p. 541-548. DOI: 10.1109/ICCAD.2017.8203824.

[48] Li B, Shan Y, Hu M, Wang Y, Chen Y, Yang H. Memristor-based approximated computation. In: International Symposium on Low Power Electronics and Design (ISLPED); 2013; Beijing, China. p. 242-247. DOI: 10.1109/ISLPED.2013.6629302.

[49] Alibart F, Zamanidoost E, Strukov DB. Pattern classification by memristive crossbar circuits using ex situ and in situ training. Nature communications. 2013 Jun 25;4(1):1-7.

[50] Chen Y, Li H, Wang X, Zhu W, Xu W, Zhang T. A nondestructive self-reference scheme for spin-transfer torque random access memory (STT-RAM). In 2010 Design, Automation & Test in Europe Conference & Exhibition (DATE 2010) 2010 Mar 8 (pp. 148-153). IEEE.

[51] Niu D, Xiao Y, Xie Y. Low power memristor-based ReRAM design with error correcting code. In 17th Asia and South Pacific Design Automation Conference 2012 Jan 12 (pp. 79-84). IEEE.

[52] Oh S, Huang Z, Shi Y, Kuzum D. The impact of resistance drift of phase change memory (PCM) synaptic devices on artificial neural network performance. IEEE Electron Device Letters. 2019 Jul 2;40(8):1325-1328.

[53] Hsieh CC, Chang YF, Jeon Y, Roy A, Shahrjerdi D, Banerjee SK. Short-Term Relaxation in HfOx/CeOx Resistive Random Access Memory with Selector. IEEE Electron Device Letters. 2017 Jun 1;38(7):871-874.

[54] Yan B, Yang J, Wu Q, Chen Y, Li H. A closed-loop design to enhance weight stability of memristor based neural network chips. In2017 IEEE/ACM International Conference on Computer-Aided Design (ICCAD) 2017 Nov 13 (pp. 541-548). IEEE.

[55] Li B, Wang Y, Chen Y, Li H, Yang H. ICE: Inline calibration for memristor crossbar-based computing engine. In: 2014 Design, Automation & Test in Europe Conference & Exhibition; 2014; Dresden, Germany. p. 1-4. DOI: 10.7873/DATE.2014.197.

[56] Liu T, Amirsoleimani A, Alibart F, Ecoffey S, Drouin D, Genov R. AIDX: Adaptive Inference Scheme to Mitigate State-Drift in Memristive VMM Accelerators. IEEE Transactions on Circuits and Systems II: Express Briefs. 2021;68:4:1128-1132. DOI: 10.1109/TCSII.2020.3026642.

[57] Kvatinsky S, Ramadan M, Friedman EG, Kolodny A, Weiser UC. TEAM: ThrEshold Adaptive Memristor Model. IEEE Transactions on Circuits and Systems I: Regular Papers. 2013;60: 1:211-221. DOI: 10.1109/TCSI.2012.2215714.

[58] Kvatinsky S, Ramadan M, Friedman EG, Kolodny A. VTEAM: A General Model for Voltage-Controlled Memristors. IEEE Transactions on Circuits and Systems II: Express Briefs. 2015;62:8:786-790. DOI: 10.1109/TCSII.2015.2433536.

[59] Biolek Z, Biolek D, Biolkova V. SPICE Model of Memristor with Nonlinear Dopant Drift. Radioengineering. 2009.

[60] Pickett MD, Strukov DB, Borghetti JL, Yang JJ, Snider GS, Stewart DR, Williams RS. Switching dynamics in titanium dioxide memristive devices. Journal of Applied Physics. 2009; 106:7:1–6. DOI: 10.1063/1.3236506

[61] Pino RE, Li H, Chen Y, Hu M, Liu B. Statistical memristor modeling and case study in neuromorphic computing. In: DAC Design Automation Conference 2012; 2012; San Francisco, CA, USA. p. 585-590. DOI: 10.1145/2228360.2228466.

[62] Lecun Y, Bottou L, Bengio Y, Haffner P. Gradient-based learning applied to document recognition. Proceedings of the IEEE. 1998; 86:11: 2278-2324. DOI: 10.1109/5.726791.

[63] Prechelt L. PROBEN 1-a set of benchmarks and benchmarking rules for neural network training algorithms. 1994.

[64] Krizhevsky A, Hinton G. Learning multiple layers of features from tiny images. 2009.

[65] Li B, Shan Y, Hu M, Wang Y, Chen Y, Yang H. Memristor-based approximated computation. In: International Symposium on Low Power Electronics and Design (ISLPED); 2013; Beijing, China. p. 242-247. DOI: 10.1109/ISLPED.2013.6629302.

[66] Strachan J, Torrezan A, Miao F, Pickett M, Yang J, Yi W, Medeiros-Ribeiro G, Williams S. State Dynamics and Modeling of Tantalum Oxide Memristors. IEEE Transactions on Electron Devices. 2013; 60:7:2194-2202. DOI: 10.1109/TED.2013.2264476

[67] Fletcher R. Practical methods of optimization. John Wiley & Sons; 2013. DOI: 10.1002/9781118723203.

Chapter 5

Pattern Formation in a RD-MCNN with Locally Active Memristors

Ahmet Samil Demirkol, Alon Ascoli, Ioannis Messaris and Ronald Tetzlaff

Abstract

This chapter presents the mathematical investigation of the emergence of static patterns in a Reaction–Diffusion Memristor Cellular Nonlinear Network (RD-MCNN) structure *via* the application of the theory of local activity. The proposed RD-MCNN has a planar grid structure, which consists of identical memristive cells, and the couplings are established in a purely resistive fashion. The single cell has a compact design being composed of a locally active memristor in parallel with a capacitor, besides the bias circuitry, namely a DC voltage source and its series resistor. We first introduce the mathematical model of the locally active memristor and then study the main characteristics of its AC equivalent circuit. Later on, we perform a stability analysis to obtain the stability criteria for the single cell. Consequently, we apply the theory of local activity to extract the parameter space associated with locally active, edge-of-chaos, and sharp-edge-of-chaos domains, performing all the necessary calculations parametrically. The corresponding parameter space domains are represented in terms of intrinsic cell characteristics such as the DC operating point, the capacitance, and the coupling resistance. Finally, we simulate the proposed RD-MCNN structure where we demonstrate the emergence of pattern formation for various values of the design parameters.

Keywords: pattern formation, memristor, reaction–diffusion, cellular nonlinear networks, destabilization, local activity, complexity

1. Introduction

An important feature of complex systems is the emergence of spatiotemporal patterns, which can be observed in numerous physical systems consisting of homogeneous media [1]. Among many examples, the emergence phenomenon can occur as a result of oscillatory kinetics in a chemical reaction [2], self-organization of biological organisms [3], mechanical vibration on liquid surfaces [4], or mineral precipitation on geologic surfaces [5]. In particular, network dynamics of pattern formation is considered to be the key attribute of information processing and memory storage in biological neural networks [6] and, hence, is the main concern in neuroscience. Following the consequences of the seminal paper of Turing [7] where he introduced the chemical basis of morphogenesis, the mechanism behind pattern formation dynamics has been extensively studied in various scientific branches. On one hand, different mathematical models have been proposed to elucidate the analytical principles of pattern formation dynamics [8]. On the other hand, there has been an ambiguity on the conceptual definition of the complexity phenomenon, which was previously described as symmetry breaking, instability of the homogenous, exchange of energy, or self-organization. In [9], Chua has proposed the theory of local activity, as the origin of complexity, where he quantitatively defined the mathematical principle behind the emergence of complex patterns in a homogenous medium. In this way, the theory of local activity assembled various definitions under the same framework and enabled the quantitative investigation of pattern formation dynamics, especially through electrical circuits. Since then, several works dealing with pattern formation dynamics on electrical hardware have referred to the theory of local activity to perform robust and quantitative analysis (e.g., see [10]). Mathematically, a well-known method to realize pattern formation dynamics is to implement reaction–diffusion partial differential equations (RD-PDEs) [11]. Various physical systems adopting reaction–diffusion equations have been shown to create well-known spatiotemporal phenomena such as traveling waves or clustering patterns [12]. Therefore, circuit implementations of RD-PDEs would be a reasonable approach to capture pattern formation dynamics on electrical hardware. Cellular nonlinear networks (CNNs), which can be described as homogeneous structures composed of evenly spaced and locally coupled identical cells, are prominent hardware solutions to implement the RD-PDEs. In particular, the reaction dynamics of an RD-PDE can be implemented by the identical cells of a CNN, while the diffusive dynamics of the same RD-PDE can be successfully discretized using the central difference approach and then implemented *via* resistive coupling between the identical neighboring cells, resulting in the so-called RD-CNN structure [13]. Motivated by this approach, emergent phenomena, such as Turing patterns or auto waves, have already been demonstrated across RD-CNNs [14].

The high-speed data transfer capability of modern mobile communication systems has led to the introduction of 5G and, in the future, 6G networks successively in a short period, which has enabled a new era in technology such as Industry 4.0 and Internet-of-Things (IoT) [15]. In parallel with the requirements of the new technological applications and the performance criteria of the corresponding hardware realizations, the design of bio-inspired neuromorphic systems utilizing the in-memory-computing principle, which stands as an alternative option to the design of conventional von-Neumann computing architectures where memory and processor units are separated from each other, has recently gained a lot of attention [16]. Similarly, there is a huge research effort in academia and industry for developing efficient fabrication techniques to implement the new in-memory computing circuit elements such as resistive switching memories, which can apparently be described as memristors [17]. The memristor, a two-terminal circuit element that was theoretically hypothesized 50 years ago, can be briefly considered as a nonlinear resistor with inherent memory dynamics [18]. Due to recent developments in physical implementations of different types of memory devices, the modeling and analysis of memristors and memristive systems have also gained attention, resulting in comprehensive circuit and system theoretical investigations [19]. Accordingly, the availability of manufactured nanoscale memristors emerges as a key enabler for the implementation of memory and processing units realized in the same place of compact hardware, and thus, provides the opportunity to design novel bio-inspired systems with in-memory-computing capabilities. In this way, it can be possible to overcome the end of Moore's law by engineering the information processing architectures, rather than downscaling the semiconductor device dimensions, which practically has come to an end.

Consistent with the above-mentioned text, locally active memristors have already been utilized in the design and application of spiking neural cells demonstrating promising results [20], while prominent artificial neural network designs employing memristor crossbar arrays have been presented in several works [21]. Besides, the theory of memristor cellular nonlinear networks (MCNNs) has been comprehensively investigated in [22–25] and the formation of Turing patterns with reaction–diffusion MCNNs (RD-MCNNs) has been introduced in [26]. In addition, pattern formation utilizing locally active NbO memristors in an MCNN has been presented in [27]. In a previous work [28], we have presented the model-based analytical investigation of dynamic pattern formation in a RD-MCNN, exploiting a manufactured locally active memristor, while the proposed single cell structure, as well as the subsequent results, is different from those presented here. Similarly, in [29], we have presented preliminary results of the mathematical investigation of static pattern formation in a RD-MCNN structure. The goal of this work is to extend the content of the previously presented works and to introduce an analytical design procedure for the implementation of static pattern formation across a compact RD-MCNN structure while taking into account the theory of local activity for the mathematical treatment. To enable the locally active dynamics, we employ a nanoscale locally active generic memristor model in the design of the basic network cell. The proposed RD-MCNN has a planar grid form and is composed of locally coupled identical cells. The compact unit cell consists of a locally active memristor in parallel with a linear capacitor, besides the bias circuitry, namely a DC voltage source and its series resistor. We first introduce the mathematical definition of the locally active generic memristor employed and, then, study the main characteristics of its AC equivalent circuit. Later on, we perform a quick stability analysis and determine stability criteria for the single cell. Consequently, we apply the theory of local activity in order to extract the parameter space with locally active, edge-of-chaos, and sharp-edge-of-chaos domains, performing all the necessary calculations parametrically. The corresponding parameter space domains are illustrated in terms of intrinsic network characteristics such as the cell DC operating point, the cell capacitance, and the coupling resistance. Essentially, we adopt a circuit theoretical approach, regarding the stability analysis of the single cell and the destabilization process after the coupling is established, which promotes an efficient investigation of the criteria to be derived. Finally, we carry out numerical simulations where we demonstrate the emergence of pattern formation across the proposed RD-MCNN structure for various values of the design parameters.

2. The locally active generic memristor

Nanoscale memristors with locally active (i.e., S-shaped) DC current–voltage (I–V) characteristics have been utilized in the design of oscillatory neuron cells [30, 31]. Furthermore, it was shown in [32–34] that memristor models in a generic form are capable of representing the dynamics of these nanoscale devices accurately. Similarly, in this work, we adopt a simplified generic memristor model to implement locally active dynamics, which helps to reduce the complexity and the simulation time of large-scale networks employing such devices. In addition, the adoption of the generic form enables simplified calculations related to the derivation of the AC model of the device, further promoting the clarification of the results.

2.1 The model definition

We introduce the memristor model equations, namely the I–V relationship in Eq. (1), and the state equation in Eq. (2), where i_m (v_m) is the memristor current (voltage), R_s is the series resistance, T is the state variable representing the temperature, and $G(T) = g_0 \bullet exp\left(-g_1/T\right)$ is the temperature-dependent memductance function. We would like to note that Eqs. (1) and (2) define a sole memristor core with current $i_{mc} = i_m$ and voltage $v_{mc} = v_m \bullet (1 + R_s \bullet G(T))^{-1}$. The parameter values of the given model employed during the numerical simulations can be found in **Table 1**.

$$i_m = v_m \bullet \frac{G(T)}{1 + R_s \bullet G(T)} = v_{mc} \bullet G(T) \tag{1}$$

$$C_T \frac{dT}{dt} = v_{mc} \bullet i_m - g_T \bullet (T - T_0) \tag{2}$$

The S-shaped DC I-V curve under the current sweep and the schematic of the complete device, which can be depicted as the series combination of the core memristor and R_s, are introduced in **Figure 1(a)**. At this point, we would like to point out that the negative differential resistance (NDR) region, which hosts the peculiar dynamics of the locally active memristor, is highlighted with the orange color in **Figure 1(a)**.

2.2 The AC equivalent circuit

Since the small-signal equivalent of the memristor plays an important role during the forthcoming circuit theoretical stability analysis, it is crucial to obtain the AC equivalent circuit of the memristor device. For this purpose, we first derive the small-signal equivalent of the core memristor and then combine it in series with R_s to obtain the AC equivalent circuit of the overall memristor. Essentially, we set $R_s = 0\Omega$ in Eqs. (1) and (2) so that the updated equation set represents the core

g_0/S	g_1/K	R_s/Ω	$C_T/(J \bullet K^{-1})$	$g_T/(W \bullet K^{-1})$	$T_0/(K)$
$5 \bullet 10^{-3}$	1700	200	10^{-14}	$6.67 \bullet 10^{-7}$	300

Table 1.
Parameter values for Eqs. (1) and (2).

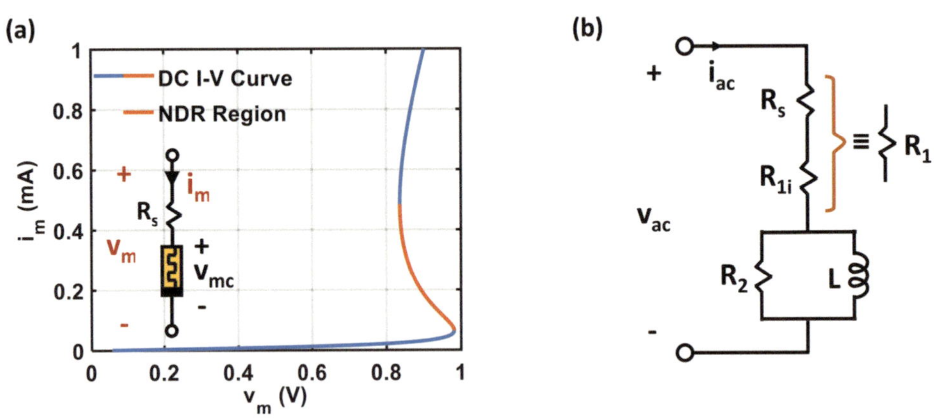

Figure 1.
(a) DC I-V curve of the locally active memristor, with NDR region highlighted. The schematic of the memristor, as the series combination between the core device and the series resistor R_s, is depicted in the inset figure. (b) AC equivalent circuit of the entire memristor where $R_1 = R_s + R_{1i}$.

memristor only. Then, we linearize Eqs. (1) and (2) by deriving their first-order Taylor series expansions. As the next step, we apply Laplace transform to the linearized equations, and after a suitable rearrangement of the Laplace equations, we express the impedance function of the core memristor in Foster's first form [35] RL circuit configuration. The final configuration of the AC equivalent circuit of the entire memristor is depicted in **Figure 1(b)**.

Regarding the AC equivalent circuit, R_{1i} represents the inverse of the slope of the DC I-V curve for $R_s = 0\Omega$ and naturally gets negative values in the NDR region. Furthermore, the quantity $(R_{1i} + R_2)$ corresponds to the instantaneous resistance, $V(T_0)/I(T_0)$, of the core memristor, which always gets positive values at a given temperature T_0, since the I-V curve lies either in the first or in the third quadrant. The same explanations are still valid also for $R_s \neq 0\Omega$ if R_{1i} is replaced by $R_1 = R_{1i} + R_s$. Lastly, L represents the dynamics of the core device, while L and R_2 have positive values at all equilibrium points. The graphical representation of the small-signal element values of **Figure 1(b)** versus the DC current I_Q can be found in **Figure 2**, where we use parameter values as given in **Table 1** during the numerical simulations. Finally, a similar procedure regarding the derivation of the AC equivalent circuit of a generic memristor including a detailed investigation and graphical illustration can be found in [36] as well.

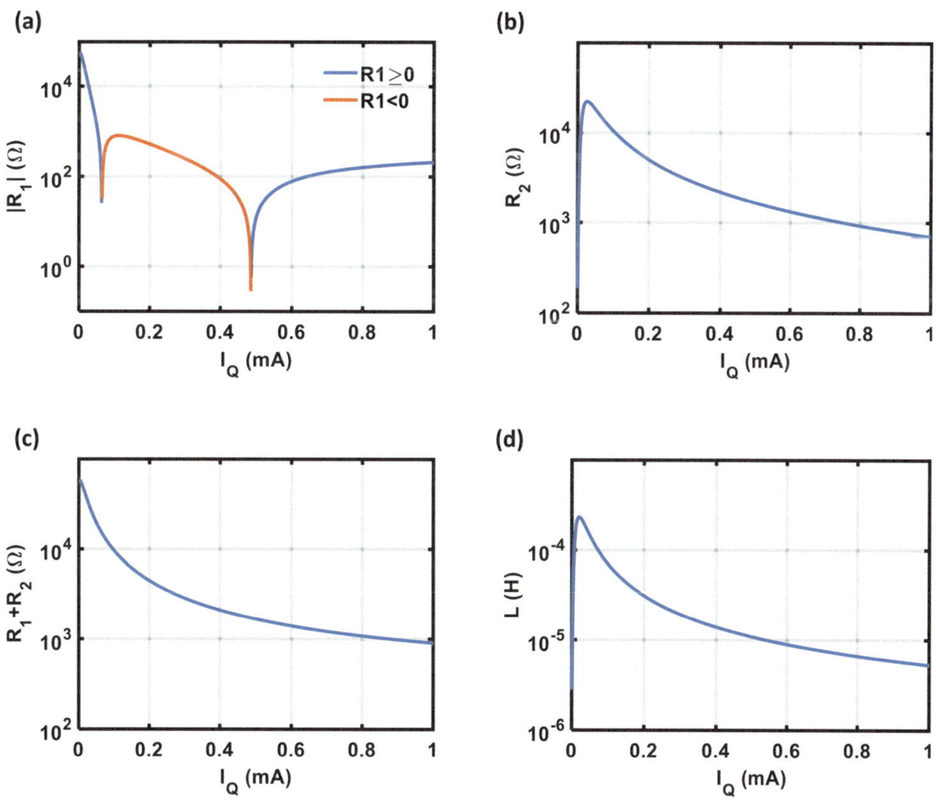

Figure 2.
Small signal elements values of ***Figure 1(b)*** *are depicted as a function of the DC equilibrium current I_Q, while the parameter values are adopted from* ***Table 1*** *during the numerical simulations. (a) $|R_1|$ vs. DC current where positive values of R_1 are depicted in blue color and negative values of R_1 are depicted in orange color. The orange part of $|R_1|$ curve one-to-one corresponds to the NDR region of the DC I-V curve. (b) R_2 vs. I_Q where R_2 always gets positive values. (c) $(R_1 + R_2)$ vs. I_Q. $(R_1 + R_2)$ corresponds to the instantaneous resistance $V(T_0)/I(T_0)$ of the memristor and always gets positive values. (d) L vs.I_Q where L represents the dynamics of the memristor and takes positive values for all current values.*

3. The single cell

To achieve a compact single cell design, a basic solution is to establish oscillatory dynamics by implementing a minimal second-order system accompanied by a simple bias circuitry. Since the memristor itself implements a first-order system with inductive dynamics, an efficient solution to increase the order of the system would be to include an additional capacitor into the cell design. Additionally, we prefer to realize the proper biasing *via* a DC voltage source and a bias resistor. Consequently, the proposed single cell is depicted in **Figure 3(a)** where V_b is the DC voltage source, R_b is the bias resistor, R_s is the series resistor, and C is the parallel capacitor, while the open circle denotes the coupling node.

3.1 Stability analysis of the single cell

Pattern formation through a RD-MCNN prerequisites stable single-cell dynamics in the isolated case. Therefore, in this section, we perform a parametric stability analysis for the uncoupled single cell, which is also the first step of parameter extraction of the circuit element values. A straightforward procedure of stability analysis would require performing linearity analysis on the state equations of the second-order single cell, followed by the derivation of the eigenvalues, to express the stability conditions [37]. In this work, we present a circuit theoretical approach where we employ the AC equivalent circuit of the single cell, which is depicted in **Figure 3(b)**, and derive the stability conditions for the uncoupled case. Adopting this approach, it is possible to express in an efficient way, the stability conditions of the isolated cell in terms of linear circuit element values and of AC element values of the memristor in a parametric form, which enables a direct evaluation of the results. Finally, although the results to be derived should hold for any equilibrium point on the DC I-V curve, without loss of generality, we assume that the memristor is biased in the NDR region, where $R_1 < 0$ and the device is locally active, which is essential for the emergence of complexity.

To derive the stability conditions of the single-cell circuit in **Figure 3(a)**, we firstly obtain the AC equivalent of it, simply by replacing the memristor with its small-signal equivalent (previously introduced in **Figure 1(b)**) and by assuming the DC voltage source as a short-circuit element, resulting in the circuit given in **Figure 3(b)**. Second, we calculate the impedance function $Z(s)$, which is illustrated in **Figure 3(b)**, and given in (3). Here, we would like to note that the poles of $Z(s)$ directly correspond to the eigenvalues of the state equations of the single cell.

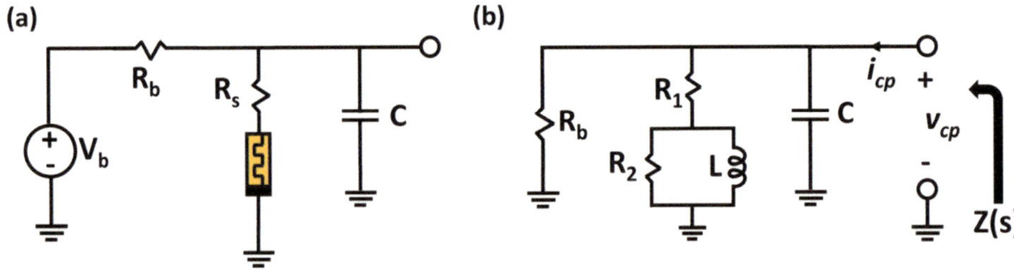

Figure 3.
(a) The single cell of the RD-MCNN structure. V_b is the DC voltage source, R_b is the bias resistor, R_s is the series resistor, C is the parallel capacitor. The open circle denotes the coupling node. (b) AC equivalent of the circuit in (a). The small-signal transfer function $Z(s)$ is the impedance seen through the coupling node.

$$Z(s) = \frac{sL(R_1+R_2)R_b + R_1R_2R_b}{s^2LC(R_1+R_2)R_b + s[L(R_1+R_2+R_b) + R_1R_2R_bC] + (R_1+R_b)R_2} = \frac{N(s)}{D(s)} \tag{3}$$

The stability of $Z(s)$ can be determined by examining the location of the poles of $Z(s)$, or equivalently, the roots of $D(s)$ via Routh-Hurwitz criteria. At this point, it is timely to recall from **Figure 2** that $L > 0,\ R_2 > 0$, and $(R_1 + R_2) > 0$, which reveals the fact that the coefficient of the s^2 term of $D(s)$ is readily positive. Consequently, to guarantee the (asymptotic) stability of $Z(s)$, remaining the coefficients of the $D(s)$ polynomial have to be nonnegative, as dictated by Routh-Hurwitz criteria. Therefore, we have $(R_1 + R_b) > 0$ and $L(R_1 + R_2 + R_b) + R_1R_2R_bC > 0$ that respectively imply that the bias resistor R_b has to be larger than the magnitude of the inverse of the slope of the DC I-V curve (i.e., $R_b > |R_1|$), while C has to be smaller than a critical C_{max} value, which is given in Eq. (4). Here, we would like to note that $D(s)$ would simplify in the absence of the parallel capacitor C, while the first stability condition $(R_1 + R_b) > 0$ would remain the same, as shown in [38], but Eq. (4) would be unnecessary.

$$C < \frac{L(R_1 + R_2 + R_b)}{-R_1R_2R_b} = C_{max} \tag{4}$$

In **Figure 4(a)**, C_{max} is shown (in blue) as a function of the equilibrium point of the memristor current I_{Q-NDR} across the entire NDR region, for two different values of R_b. It can be seen from **Figure 4(a)** that the smaller value of R_b results in a larger value for C_{max}, which would be beneficial during a hardware realization, while we remind that R_b should be kept larger than $|R_1|$ for a stable operation.

3.2 Local activity analysis of the single cell

The small-signal behavior of an uncoupled cell, which is biased at a locally passive operating point, can be represented by a positive real complexity function (or similarly, by a positive real transfer function for the 1-port coupling case). Subsequently, the AC equivalent circuit of the same cell at the given operating point would be strictly composed of passive linear elements, which inherently results in stable dynamics. Therefore, a diffusive coupling (i.e., the coupling established *via* a resistor) obtained between these identical cells would similarly result in an

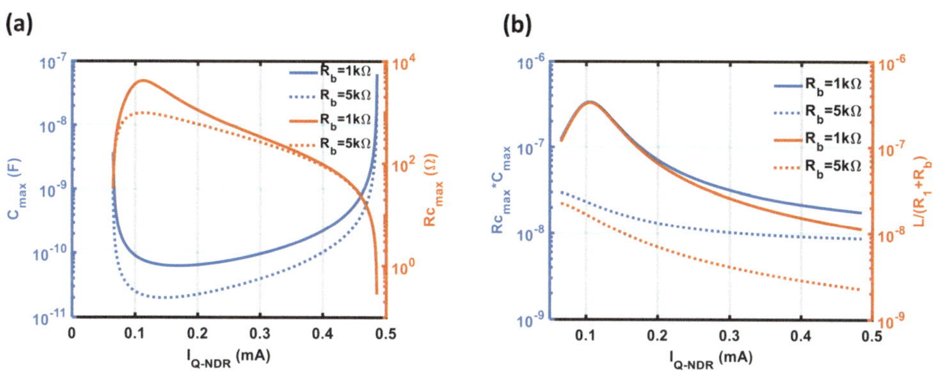

Figure 4.
*(a) C_{max} vs. I_{Q-NDR} (blue curves) and R_{c-max} vs. I_{Q-NDR} (orange curves). (b) C_{max} * R_{c-max} product. It can be seen from both graphs that smaller values of R_b results in larger values for C_{max}, R_{c-max}, and C_{max} * R_{c-max} values, while the last product term is more influenced from R_b value.*

eventually stable structure since the combination of passive circuit elements through a resistive coupling cannot give rise to instability. Thus, to be able to observe the emergence of complexity in a diffusively coupled homogenous network, it is vital to accommodate locally active cells, which cannot be represented by positive real complexity functions in the small-signal regime. This fact at the same time implies that local activity can be simply judged as a violation of local passivity [9]. Furthermore, a locally active cell is defined to be on the edge-of-chaos, if it is biased at an asymptotically stable operating point. Most importantly, an uncoupled cell, which is poised on the edge-of-chaos and therefore considered as a "dead" or "silent" cell due to its stability, can be potentially destabilized from its quite state *via* resistive coupling, which leads to the generation of complex patterns across the homogenous medium. Correspondingly, the edge-of-chaos domain involves a subset called the sharp-edge-of-chaos domain, which defines the set of parameters that destabilize the cell after coupling is introduced. Strictly speaking, a one-port cell is said to be locally active if and only if its small-signal transfer function (e.g., $Z(s)$ given by Eq. (3) in our case) satisfies any of the conditions below [9]:

1. $Z(s)$ has a pole with positive real part, that is, $Re\,[s] > 0$.

2. $Z(s)$ has multiple poles on the *imaginary* ($j\omega$) axis.

3. $Z(s)$ has a simple pole $s = j\omega_p$ on the *imaginary* axis and the residue associated with this pole, specifically $r(j\omega_p) = \lim_{s \to j\omega_p} Z(s) \bullet (s - j\omega_p)$, is either a negative real number or a complex number.

4. $Re\,[Z(j\omega)] < 0$ for some $\omega \in (-\infty, \infty)$.

To confirm the locally active dynamics of the complexity function $Z(s)$ of Eq. (3), which is associated with the uncoupled cell of **Figure 3(a)**, a direct approach would be to check whether these four rigorously defined criteria are satisfied or not, an approach we had applied in a previous work [28]. However, in this work, we apply a quick inspection method of local activity criteria, and rather check if $Z(s)$ clearly violates local passivity. In consistent with this approach, it can be seen that $Z(s)$ possesses a right half plane (RHP) zero (i.e., $z = -R_1R_2/L(R_1 + R_2) > 0$, for $R_1 < 0$), a feature that cannot be realized with any locally passive transfer function. Thus, without a need for a further examination, we can infer that once it is biased in the NDR region, the uncoupled cell operates in the locally active regime. Furthermore, since the uncoupled cell is designed to be asymptotically stable (*via* tuning R_b and C accordingly), we can directly conclude that it is both locally active and on the edge-of-chaos across the entire NDR region. Finally, we would like to note that the RHP zero is a strong indicator of a possible destabilization scenario after the coupling is established.

3.3 Destabilization analysis after the introduction of coupling

For the emergence of pattern formation, it is essential that the identical cells lose stability after the resistive coupling is established between them. In the previous section, we have studied the stability of the uncoupled cell, while in this section, we investigate the destabilization process after a resistive array is added to couple the elements of the network. Since a direct stability analysis that requires exploring the complete network itself would be complicated and time-consuming, it would be beneficial to apply a simplified stability analysis. Considering highly accurate illustrative examples, the stability analysis of two-cell [37] and three-cell [29] arrays can be found in the literature. In this work, we present a quick inspection method, which is conceptually introduced in [9] and has been implemented through an example in [28], where the idea is to terminate the coupling node of the single cell with a coupling resistor, resulting in the circuit depicted in **Figure 5(a)** for the case study presented here.

To investigate the possible destabilization scenarios of the resistively terminated cell in **Figure 5(a)**, we follow the same circuit theoretical approach as in the previous subsections and examine the corresponding AC equivalent circuit, which is shown in **Figure 5(b)**. Here, the poles of the impedance function $Z_c(s)$ coincide with the eigenvalues of the second-order system equivalent of the circuit in **Figure 5(a)** at a given equilibrium point and should be examined to determine the destabilization conditions. At this point, it should be noted that the AC equivalent circuits depicted in **Figures 3(b)** and **5(b)** are qualitatively equivalent to each other, while R_c appears additionally in parallel with R_b in **Figure 5(b)**. Therefore, the expression of $Z_c(s)$, readily given in Eq. (5), is identical to the expression of $Z(s)$, where the term R_b is replaced with the parallel equivalent term R_{bc}, that is, $R_{bc} = R_b \bullet R_c/(R_b + R_c)$.

$$Z_c(s) = \frac{sL(R_1 + R_2)R_{bc} + R_1R_2R_{bc}}{s^2LC(R_1 + R_2)R_{bc} + s[L(R_1 + R_2 + R_{bc}) + R_1R_2R_{bc}C] + (R_1 + R_{bc})R_2} = \frac{N_c(s)}{D_c(s)} \tag{5}$$

To claim instability of $Z_c(s)$, at least one coefficient of the $D_c(s)$ polynomial has to be negative, as dictated by Routh-Hurwitz criteria. Here, it is straightforward to see that the coefficient of the s^2 term of $D_c(s)$ is readily positive. Similarly, it is possible to show after some algebraic rearrangement that the coefficient of the s term remains positive as long as the stability precondition given by Eq. (4), with R_b replaced by R_{bc}, is satisfied. Thus, for instability, the constant term of $D_c(s)$, namely $(R_1 + R_{bc}) \bullet R_2$, has to be negative. Starting with the inequality $(R_1 + R_{bc}) \bullet R_2 < 0$ and replacing R_{bc} with $R_b \bullet R_c/(R_b + R_c)$ directly give us the destabilization condition introduced by Eq. (6).

$$R_c < -\frac{R_1 \bullet R_b}{R_1 + R_b} = R_{c-max} = -(R_1 \| R_b) \tag{6}$$

In **Figure 4(a)**, we plot (in orange) R_{c-max} as a function of the equilibrium point of the memristor current I_{Q-NDR} across the entire NDR region, for two different values of R_b. Similar to the characteristics of C_{max}, the smaller value of R_b results in a larger value for R_{c-max}, which would relieve the design constraints for an

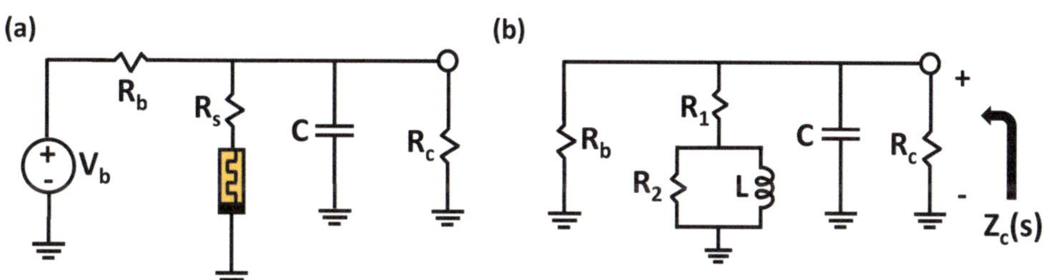

Figure 5.
(a) A simplified scenario for the resistively coupled single cell where R_c stands for the coupling resistor. (b) AC equivalent circuit of the cell in (a). The poles of $Z_c(s)$ shall be investigated for analyzing the stability of the circuit in (a) at a given equilibrium point.

hardware realization. On the other hand, Eq. (6) reveals the fact that considering the memristor model-dependent quantities, R_{c-max} depends only on the slope of the I-V curve (i.e., R_1^{-1}), while C_{max} depends on other quantities (see Eq. (4)) as well. Besides, as a function of I_{Q-NDR}, the characteristics of C_{max} increase (decrements), while the characteristics of R_{c-max} decrease (increments). Moreover, it is interesting to mention that the condition derived in Eq. (6) is qualitatively equivalent to the results presented in [29, 37], indicating the fact that the simplification considered in this work is accurate and provides a quick insight into the destabilization scenario of the coupled cell in **Figure 5(a)**.

Considering a typical time constant based design approach, we characterize the term $R_{c-max} \bullet C_{max}$, introducing the exact and an approximate product in Eq. (7), and plot these quantities in **Figure 4(b)** as a function of I_{Q-NDR}, for two different values of R_b. Here, we would like to note that the approximate product term $L/(R_1 + R_b)$ is obtained under the assumption $(R_1 + R_2 + R_b) \cong R_2$.

$$R_{c-max} \bullet C_{max} = -\frac{R_1 \bullet R_b}{R_1 + R_b} \bullet \frac{L(R_1 + R_2 + R_b)}{-R_1 R_2 R_b} = \frac{L(R_1 + R_2 + R_b)}{R_2(R_1 + R_b)} \cong \frac{L}{(R_1 + R_b)} \tag{7}$$

It can be seen from **Figure 4(b)** that the approximated time constant term given in Eq. (7) follows the original expression very closely, especially for the smaller value of R_b. Moreover, the approximated value is independent of R_2, but rather dependent on the dynamic quantity L, or in other words, the switching speed of the memristor device. As a final and crucial remark, referring to the instability region of parameters, that is, the sharp-edge-of-chaos domain, it is possible to rigorously define the necessary circuit design variables by taking into account the parametric equations given by Eqs. (4) and (6).

4. RD-MCNN structure and static pattern formation

The proposed RD-MCNN structure has a planar grid arrangement and it is composed of $m \times n$ identical cells where m is the number of rows, n is the number of columns, and $C_{i,j}$ represents the cell at i^{th} row and the jth column. All the cells are resistively coupled to the respective nearest neighbors in vertical and horizontal directions only, and the CNN structure assumes periodic boundary conditions; that is, the first and the last cells in each row, and respectively, in each column, are resistively coupled to each other as well. To study the emergence of pattern formation dynamics in the proposed RD-MCNN, among several structures investigated, here we present typical simulation results of a two-dimensional 51×51 structure. We define the same initial conditions ($T_0 = 300K$ and $V_{c0} = 0V$) for all the cells unless otherwise stated, except for the cell, namely $C_{26,26}$, which is in the center of the network. This exception for the definition of initial conditions is adopted to initiate the emergence of the transient behavior in the numerical simulations.

First of all, we investigate the effect of changing the DC operating point on the static pattern formation. To this end, we set the design parameters R_b to $2k\Omega$ and R_c to $0.1k\Omega$, while we tune V_b to vary the location of the DC operating point of each cell. We present the emergent static patterns in **Figure 6**, where $V_b = 1.2V$ in **Figure 6(a)**, $V_b = 1.3V$ in **Figure 6(b)**, $V_b = 1.4V$ in **Figure 6(c)**, and $V_b = 1.5V$ in **Figure 6(d)**. Here, we plot the memristor voltage, or equivalently, the capacitor voltage, of each cell and assign a color code to its amplitude value, as depicted on the right side of each plot. It is possible to observe from **Figure 6** that a shift in the location of the operating point due to an increase in V_b may lead to static patterns featuring a reduced spread in capacitor voltage amplitude through the array,

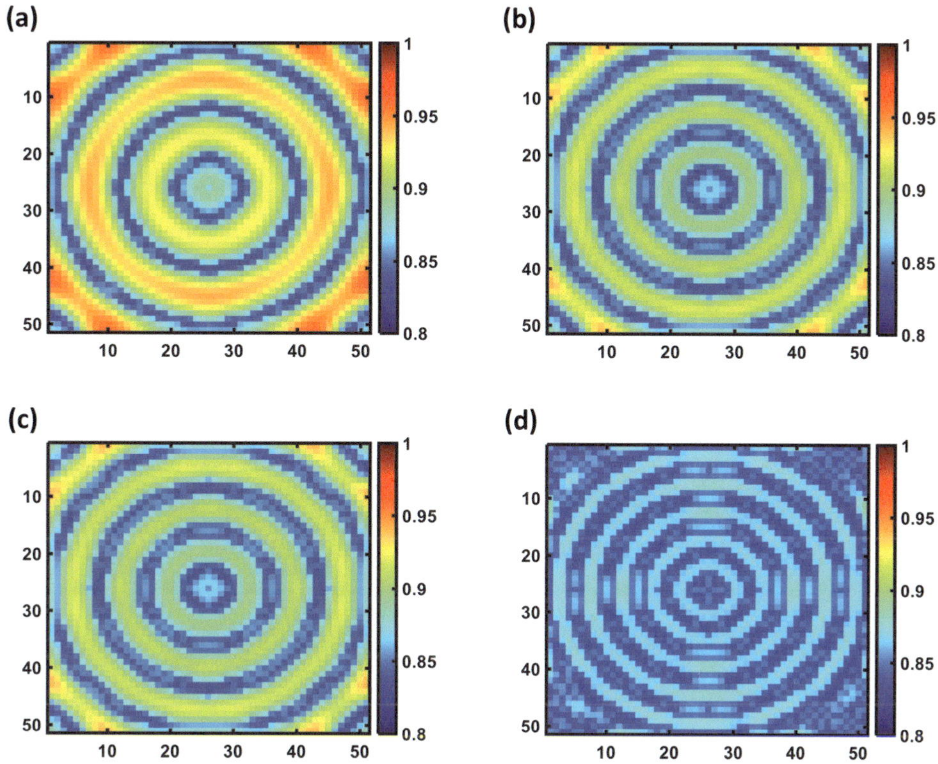

Figure 6.
RD-MCNN output values represented by the cell memristor voltages obtained in a simulation of a two-dimensional 51 × 51 RD-MCNN structure for $R_b = 2k\Omega$, $R_c = 0.1k\Omega$, while $V_b = 1.2V$ in (a), $V_b = 1.3V$ in (b), $V_b = 1.4V$ in (c), and $V_b = 1.5V$ in (d). All the cells have the same initial conditions except for the center cell $C_{26,26}$. A shift in the memristor DC operating point may lead to static patterns with a reduced range in capacitor voltage amplitude, as mostly pronounced in (d).

reflected by a gradual decrement in the variety of colors in the emergent patterns, as especially observed in **Figure 6(d)**.

Later on, we examine the effect of the coupling strength on the static pattern formation by varying the value of the coupling resistance R_c. Similarly, we set the design parameters V_b to 1.2V and R_b to 2$k\Omega$ while we tune the value of R_c to adjust the coupling strength. The results are illustrated in **Figure 7**, where $R_c = 0.25k\Omega$ in **Figure 7(a)**, $R_c = 0.5k\Omega$ in **Figure 7(b)**, $R_c = 1k\Omega$ in **Figure 7(c)**, and $R_c = 2k\Omega$ in **Figure 7(d)**. It can be seen that as R_c increases, first a deformation starts to occur in the outer parts of the static pattern of **Figure 7(a)–(c)**, while clearly, a new static pattern emerges, finally in **Figure 7(d)**. The final pattern in **Figure 7(d)** also shows that neighboring cells exhibit a sharper spread in the capacitor voltage amplitude, which is reflected by the clear color contrast observable in this plot, especially as compared to that of **Figure 7(a)**.

In addition, we focus on the effect of the initial conditions on pattern formation, as depicted in **Figure 8**, where we fix the values of all of the design parameters such that $V_b = 1.2V$, $R_b = 2k\Omega$, and $R_c = 0.1k\Omega$. The cells located in the center of each side of the edges (i.e., $C_{1,26}$, $C_{26,1}$, $C_{51,26}$, $C_{26,51}$), or midpoint cells for short, have the same initial condition as the center cell $C_{26,26}$ in **Figure 8(a)**, while only the corner cells ($C_{1,1}$, $C_{1,51}$, $C_{51,1}$, $C_{51,51}$) have the same initial condition as the center cell $C_{26,26}$ in **Figure 8(c)**. On the other hand, only the midpoint cells share the same initial conditions in **Figure 8(b)**, while the center cell $C_{26,26}$ features the same initial condition as the rest of the network. Similarly, only the corner cells share the same

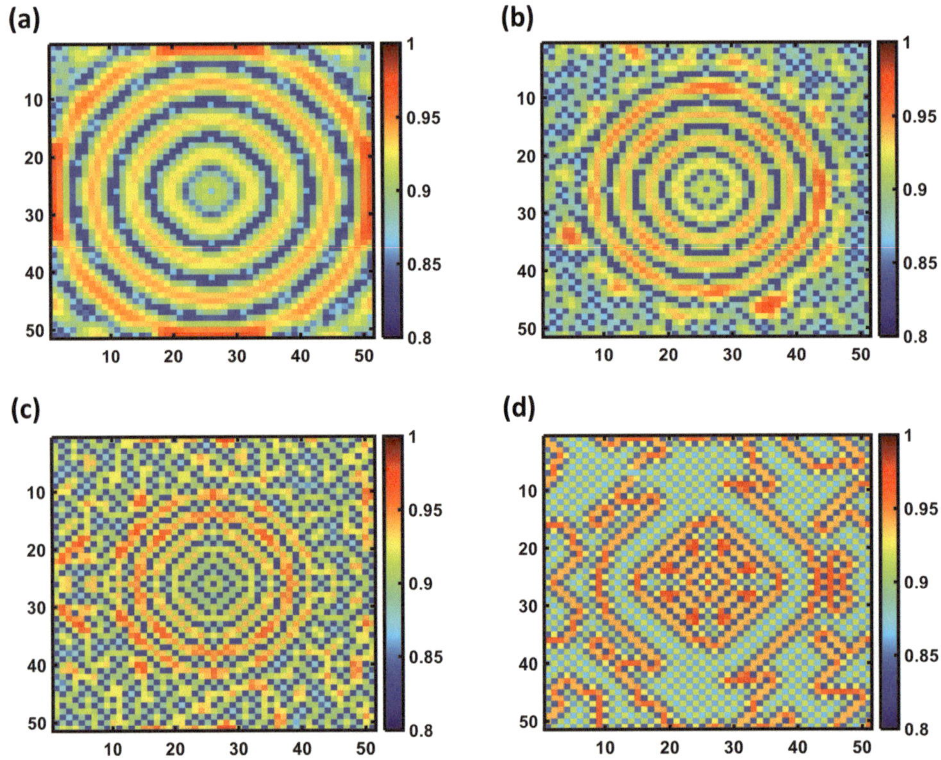

Figure 7.
RD-MCNN output values represented by the cell memristor voltages obtained in a simulation of a two-dimensional 51 × 51 RD-MCNN structure for $R_b = 2k\Omega$, $V_b = 1.2V$, while $R_c = 0.25k\Omega$ in (a), $R_c = 0.5k\Omega$ in (b), $R_c = 1k\Omega$ in (c), and $R_c = 2k\Omega$ in (d). All the cells have the same initial condition except for the center cell $C_{26,26}$. As R_c increases, first a deformation occurs in the outer parts of the pattern observed in (a), as shown in (b) and (c), while a clearly new pattern emerges in (d). With reference to the pattern in (d), the capacitor voltages in neighboring cells exhibit a sharper change in the amplitude, which is practically reflected by the color contrast observable, especially as compared to that of (a).

initial conditions in **Figure 8(d)**, while the center cell $C_{26,26}$ features the same initial condition as the rest of the network cells. The difference between the patterns presented in **Figure 8(a)** and **(b)**, and equivalently in **Figure 8(c)** and **(d)**, shows the effect of a change in the number of cells sharing the same initial conditions. Likewise, the difference between the patterns presented in **Figure 8(a)** and **(c)**, and equivalently in **Figure 8(b)** and **(d)**, shows the effect of a change in the location (indeed, a rotation of half-side length) of cells sharing the same initial conditions, which can result in clearly different patterns. Since there exists a very high number of spatial permutations for the location and the number of cells with the same initial conditions, we conjecture that there may appear a large class of clearly distinguishable patterns, such as those presented in **Figure 8**, which results in a significant memory capacity of the network.

Lastly, we investigate the effect of the size of the network on the patterns generated through. For this purpose, we set the design parameters V_b to $1.2V$, R_b to $2k\Omega$, and R_c to $0.2k\Omega$, while we define the same initial conditions for the corner cells as it is the case for the center cell. Then, we simulate structures of different size while preserving their square geometry, and depict the results in **Figure 9**, where $m = n = 31$ in **Figure 9(a)**, $m = n = 41$ in **Figure 9(b)**, $m = n = 51$ in **Figure 9(c)**, and $m = n = 61$ in **Figure 9(d)**. Once more, it can be concluded from **Figure 9** that the patterns generated across networks of different sizes and square geometry are clearly distinguishable one from the other.

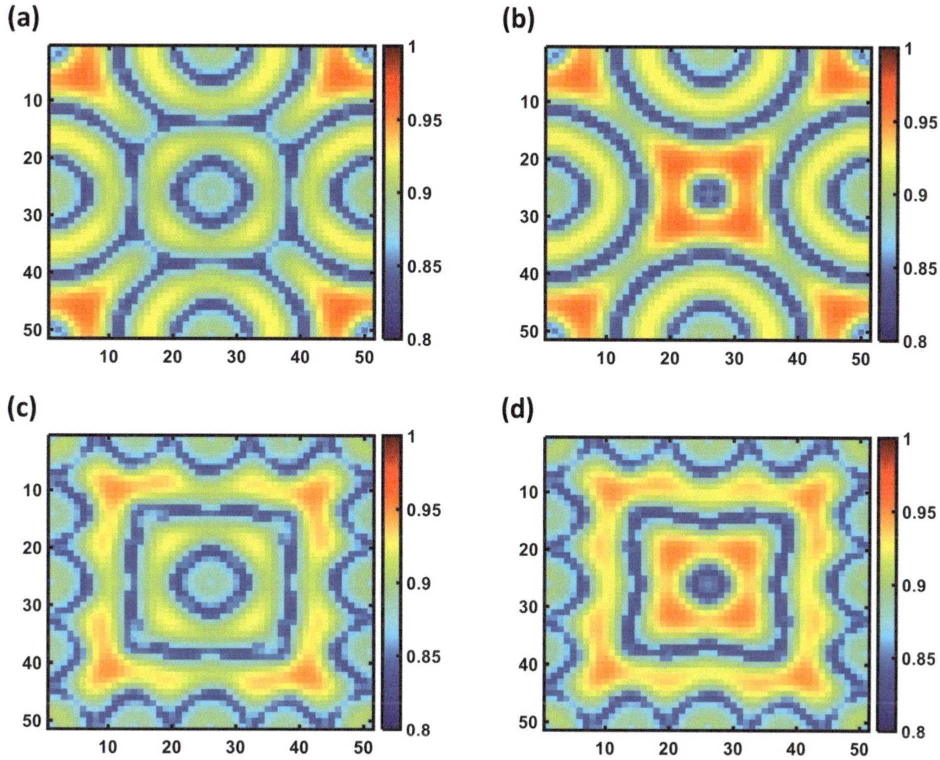

Figure 8.
RD-MCNN output values represented by the cell memristor voltages obtained in a simulation of a two-dimensional 51×51 RD-MCNN structure for $R_b = 2k\Omega$, $R_c = 0.1k\Omega$ and $V_b = 1.2V$. The cells in the middle of each side of the edges, or briefly the midpoint cells ($C_{1,26}$, $C_{26,1}$, $C_{51,26}$, $C_{26,51}$), have the same initial conditions as the center cell $C_{26,26}$ in (a), while the corner cells ($C_{1,1}$, $C_{1,51}$, $C_{51,1}$, $C_{51,51}$) have the same initial conditions as the center cell $C_{26,26}$ in (c). On the other hand, only the midpoint cells share the same initial conditions in (b), while the center cell $C_{26,26}$ features the same initial condition as the rest of the network cells. Similarly, only corner cells share the same initial conditions in (d), while the center cell $C_{26,26}$ features the same initial conditions as the rest of the network cells. A change in the location as well as in the number of cells sharing the same initial conditions can result in clearly different patterns, conferring the network a significant memory capacity.

5. Conclusion

In this chapter, we have presented the mathematical investigation of static pattern formation across a resistively coupled RD-MCNN structure *via* the application of the theory of local activity. The considered networks have identical cells in a compact form, each of which is composed of a DC bias voltage source, a bias resistor, a locally active memristor, and a parallel capacitor. The memristor was represented through a generic model, which helped to reduce the numerical complexity and to shorten the simulation time of large-scale networks. A useful AC equivalent circuit could be derived for the locally active device, which facilitated further calculations related to the small signal model of the network cell. We have adopted a systematic circuit theoretical approach that we applied for the stability analysis of the isolated cell and for the extraction of its parameter values for its operation on the edge-of-chaos and sharp-edge-of-chaos domains. In this way, we have performed a simple, fast, and robust analysis, which at the same time allowed an efficient interpretation of the results. All the calculations were performed in parametric form, which allowed a deep investigation of the results, making it possible to extract the related sets of parameters in terms of the cell characteristics, namely the DC operating point and the capacitor value, as well as the value of the

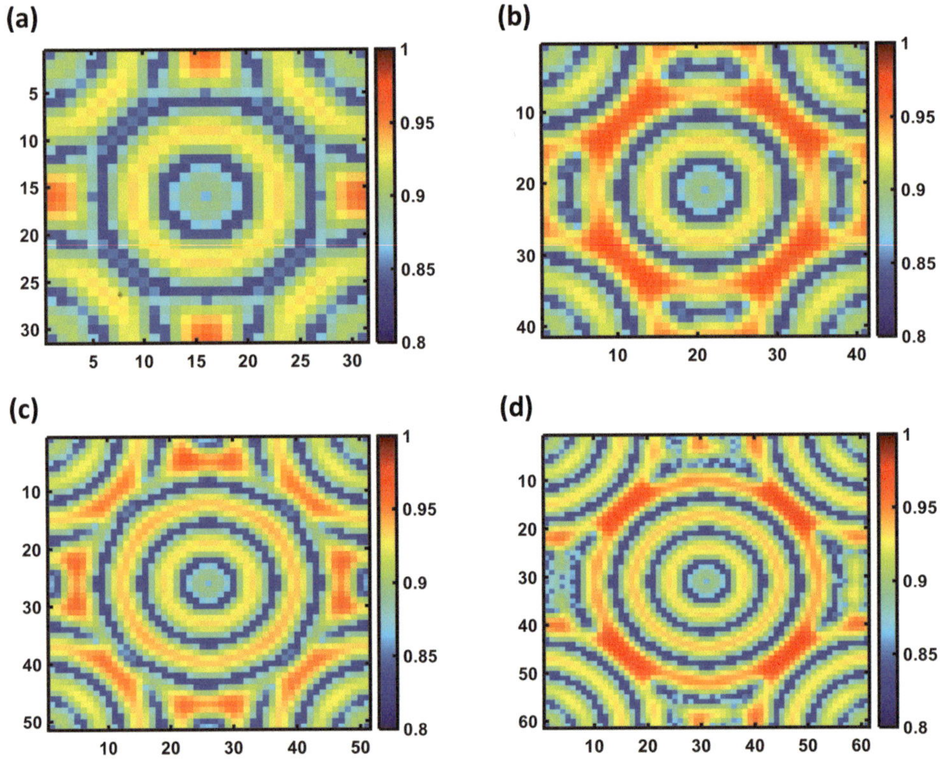

Figure 9.
Simulation results of various $m \times n$ RD-MCNN structures for $R_b = 2k\Omega$, $R_c = 0.2k\Omega$ and $V_b = 1.2V$, where $m = n = 31$ in (a), $m = n = 41$ in (b), $m = n = 51$ in (c), and $m = n = 61$ in (d). The corner cells have the same initial conditions as the center cell for all the structures. Although the four structures share a square geometry, they differ in size, which leads to clearly distinguishable patterns.

coupling resistance. The proposed RD-MCNN was shown to generate diverse patterns while we have extensively investigated the effect of design parameters, initial conditions, and the size of the network on the patterns generated therein. Furthermore, the mathematical and circuit theoretical approach employed during the investigation of pattern formation dynamics in the RD-MCNN under focus in this chapter can be easily adopted to RD-MCNNs with different cell designs.

Finally, the content of this work can be extended to the investigation of the role of the network geometry and variety in cells' initial conditions on pattern formation. Furthermore, the impact of memristor non-idealities such as variability in memristor behavior from cycle to cycle, as well as from device to device, and endurance degradation and short- and long-term reliability issues shall be further examined to achieve a robust design. Such non-idealities, essentially, can affect the NDR characteristics of the memristors and narrow the width of the NDR region, or the devices can even get stuck in one of the high-resistance or low-resistance locally passive regimes where they act as dead cells without any dynamics. Preliminary simulation results show that endowing memristors with narrower NDR width negatively affect the patterns' color contrasts, while the presence of the high- or low-resistance locally passive devices results in the formation of color clusters within the patterns. Interestingly, similar color clusters emerge also when some of the array memristors' initial conditions are randomly selected, which suggests a possible strategy to compensate for the disturbing action of locally passive cells on the formation of predefined patterns through a dynamic conditioning of their initial conditions. Similarly, the low color contrast quality in the patterns, which occur due to the existence of memristors featuring a narrower NDR region, can be improved by reprogramming dynamically the bias voltage sources in these "defect" cells. In conclusion, proper control strategies including cells' dynamic biasing and initial condition conditioning, as well as reconfiguration of the network array, may be considered as a solution to overcome the harmful effects which memristor non-idealities may induce on the emergence of predefined patterns in the proposed RD-MCNNs. Future studies will also be devoted to envision an application, where the capability of our cellular medium to generate a variety of steady-state static patterns may be useful to our modern society.

Acknowledgements

We kindly acknowledge the support for the Book Processing Charge by the Open Access Publication Fund of Saxon State and University Library Dresden (SLUB). This work was also supported by the German Research Foundation (DFG) under Project No. 411647366.

Author details

Ahmet Samil Demirkol*, Alon Ascoli, Ioannis Messaris and Ronald Tetzlaff
Technische Universität Dresden (TUD), Dresden, Germany

*Address all correspondence to: ahmet_samil.demirkol@tu-dresden.de

References

[1] Walgraef D. Spatio-Temporal Pattern Formation with Examples from Physics, Chemistry, and Materials Science. NY: Springer-Verlag; 1997. p. 306. DOI: 10.1007/978-1-4612-1850-0

[2] Maini PK, Painter KJ, Chau HNP. Spatial pattern formation in chemical and biological systems. Journal of the Chemical Society, Faraday Transactions. 1997;**93**(20):3601-3610. DOI: 10.1039/A702602A

[3] Demir E, Yaman IY, Basaran M, Kocabas A. Dynamics of pattern formation and emergence of swarming in *Caenorhabditis elegans*. eLife. 2020;**9**: 1-21. DOI: 10.7554/eLife.52781

[4] Shao X, Wilson P, Saylor J, Bostwick J. Surface wave pattern formation in a cylindrical container. Journal of Fluid Mechanics. 2021;**915**: A19. DOI: 10.1017/jfm.2021.97

[5] Meakin P, Jamtveit B. Geological pattern formation by growth and dissolution in aqueous systems. Proceedings of the Royal Society A. 2010;**466**:659-694. DOI: 10.1098/rspa.2009.0189

[6] Hutt MT, Kaiser M, Hilgetag CC. Perspective: Network-guided pattern formation of neural dynamics. Philosophical Transactions of the Royal Society B. 2014;**369**:20130522. DOI: 10.1098/rstb.2013.0522

[7] Turing AM. The chemical basis of morphogenesis. Philosophical Transactions. Royal Society of London. 1952;**B237**:37-72. DOI: 10.1098/rstb.1952.0012

[8] Sekimura T, Noji S, Ueno N, Maini PK, editors. Morphogenesis and Pattern Formation in Biological Systems. Japan: Springer; 2003. p. 398. DOI: 10.1007/978-4-431-65958-7

[9] Chua L. Local activity is the origin of complexity. International Journal of Bifurcation and Chaos. 2005;**15**(11): 3435-3456. DOI: 10.1142/S0218127405014337

[10] Slavova A, Tetzlaff R. Mathematical analysis of memristor CNN, memristors. In: James A, editor. Circuits and Applications of Memristor Devices. IntechOpen. 2019. DOI: 10.5772/intechopen.86446

[11] Gilli M, Roska T, Chua L, Civalleri PP. On the relationship between CNNs and PDEs. In: Proc. of the 7th IEEE Int. Workshop on Cellular Neural Networks and Their App. 24-24 July 2002; Germany: IEEE; 2002. pp. 16-24. DOI:10.1109/CNNA.2002.1035030

[12] Arena P, Fortuna L, Branciforte M. Reaction-diffusion CNN algorithms to generate and control artificial locomotion. IEEE Transactions on Circuits and Systems I: Fundamental Theory and Applications. 1999;**46**(2): 253-260. DOI: 10.1109/81.747195

[13] Chua L. CNN: A Paradigm for Complexity. World Scientific Series on Nonlinear Science Series A. Singapore: World Scientific Publishing Co. Pte. Ltd.; 1998. Vol. 31. p. 332. DOI: 10.1142/3801

[14] Buscarino A, Corradino C, Fortuna L, Frasca M. Turing patterns via pinning control in the simplest memristive cellular nonlinear networks. Chaos: An Interdisciplinary Journal of Nonlinear Science. 2019;**29**:103145. DOI: 10.1063/1.5115131

[15] 5G Evolution and 6G [Internet]. Available from: https://www.nttdocomo.co.jp/english/binary/pdf/corporate/technology/whitepaper_6g/DOCOMO_6G_White_PaperEN_20200124.pdf [Accessed: 01 August 2021]

[16] Ielmini D, Wong HSP. In-memory computing with resistive switching devices. Nature Electronics. 2018;**1**: 333-343. DOI: 10.1038/s41928-018-0092-2

[17] Slesazeck S, Mikolajick T. Nanoscale resistive switching memory devices: A review. Nanotechnology. 2019;**30**: 352003. DOI: 10.1088/1361-6528/ab2084

[18] Chua L. Memristor—The missing circuit element. IEEE Transactions on Circuit Theory. 1971;**18**(5):507 519. DOI: 10.1109/TCT.1971.1083337

[19] Chua L. Everything you wish to know about memristors but are afraid to ask. Radioengineering. 2015;**24**(2): 319-368. DOI: 10.13164/re.2015.0319

[20] Pickett M, Medeiros-Ribeiro G, Williams RS. A scalable neuristor built with Mott memristors. Nature Materials. 2013;**12**:114-117. DOI: 10.1038/nmat3510

[21] Xia Q, Yang JJ. Memristive crossbar arrays for brain-inspired computing. Nature Materials. 2019;**18**:309-323. DOI: 10.1038/s41563-019-0291-x

[22] Tetzlaff R, Ascoli A, Messaris I, Chua L. Theoretical foundations of memristor cellular nonlinear networks: Memcomputing with bistable-like memristors. IEEE Transactions on Circuits and Systems I: Regular Papers. 2020;**67**(2):502-515. DOI: 10.1109/TCSI.2019.2940909

[23] Ascoli A, Messaris I, Tetzlaff R, Chua L. Theoretical foundations of memristor cellular nonlinear networks: Stability analysis with dynamic memristors. IEEE Transactions on Circuits and Systems I: Regular Papers. 2020;**67**(4):1389-1401. DOI: 10.1109/TCSI.2019.2957813

[24] Ascoli A, Tetzlaff R, Kang SM, Chua L. Theoretical foundations of memristor cellular nonlinear networks: A DRM2-based method to design memcomputers with dynamic memristors. IEEE Transactions on Circuits and Systems I: Regular Papers. 2020;**67**(8):2753-2766. DOI: 10.1109/TCSI.2020.2978460

[25] Ascoli A, Tetzlaff R, Kang SM, Chua L. System-theoretic methods for designing bio-inspired mem-computing memristor cellular nonlinear networks. Frontiers in Nanotechnology. 2021;**3**: 633026. DOI: 10.3389/fnano.2021.633026

[26] Buscarino A, Corradino C, Fortuna L, Frasca M, Chua L. Turing patterns in memristive cellular nonlinear networks. IEEE Transactions on Circuits and Systems I: Regular Papers. 2016;**63**(8):1222-1230. DOI: 10.1109/TCSI.2016.2564738

[27] Weiher M, Herzig M, Tetzlaff R, Ascoli A, Mikolajick T, Slesazeck S. Pattern formation with locally active S-type NbOx memristors. IEEE Transactions on Circuits and Systems I: Regular Papers. 2019;**66**(7):2627-2638. DOI: 10.1109/TCSI.2019.2894218

[28] Demirkol AS, Ascoli A, Messaris I, Tetzlaff R. Analytical investigation of pattern formation in an M-CNN with locally active NbOx memristors. In: IEEE International Symposium on Circuits and Systems (ISCAS2021). 2021. pp. 1-5. DOI:10.1109/ISCAS51556.2021.9401280

[29] Demirkol AS, Ascoli A, Tetzlaff R. Mathematical investigation of static pattern formation with a locally active memristor model. In: The 17th IEEE International Workshop on Cellular Nanoscale Networks and their Applications. Catania: CNNA2021 [Accepted for Publication]

[30] Kumar S, Strachan JP, Williams RS. Chaotic dynamics in nanoscale NbO2 Mott memristors for analogue computing. Nature. 2017;**548**:318-321. DOI: 10.1038/nature23307

[31] Herzig M, Weiher M, Ascoli A, Tetzlaff R, Mikolajick T, Slesazeck S. Multiple slopes in the negative differential resistance region of NbOx-based threshold switches. Journal of Physics D: Applied Physics. 2019;**52**: 325104. DOI: 10.1088/1361-6463/ab217a

[32] Messaris I, Tetzlaff R, Ascoli A, Williams RS, Kumar S, Chua L. A simplified model for a NbO2 Mott memristor physical realization. In: IEEE International Symposium on Circuits and Systems (ISCAS); 2020. pp. 1-5. DOI:10.1109/ISCAS45731.2020.9181036

[33] Ascoli A, Demirkol AS, Tetzlaff R, Slesazeck S, Mikolajick T, Chua L. On local activity and edge of chaos in a NaMLab memristor. Frontiers in Neuroscience. 2021;**15**:651452. DOI: 10.3389/fnins.2021.651452

[34] Li S, Liu X, Nandi SK, Nath SK, Elliman RG. Origin of current-controlled negative differential resistance modes and the emergence of composite characteristics with high complexity. Advanced Functional Materials. 2019;**29**(44):1905060. DOI: 10.1002/adfm.201905060

[35] Kuo FF. Network Analysis and Synthesis. Wiley International Edition, Singapore: John Wiley & Sons; 1966. p. 515

[36] Messaris I, Brown T, Demirkol AS, Al Chawa MM, Williams RS, Tetzlaff R, et al. NbO2-Mott memristor: A circuit-theoretic investigation. IEEE Transactions on Circuits and Systems I: Regular Papers. 2021 [Under Review]

[37] Ascoli A, Demirkol AS, Tetzlaff R, Chua L. Edge of chaos theory resolves smale paradox in the simplest memristor oscillatory network. IEEE Transactions on Circuits and Systems I: Regular Papers. 2021 [Under Review]

[38] Ascoli A, Slesazeck S, Mähne H, Tetzlaff R, Mikolajick T. Nonlinear dynamics of a locally-active memristor. IEEE Transactions on Circuits and Systems I: Regular Papers. 2015;**62**(4): 1165-1174. DOI: 10.1109/TCSI.2015.2413152

Chapter 6

Development of Compute-in-Memory Memristive Crossbar Architecture with Composite Memory Cells

Mehri Teimoory, Amirali Amirsoleimani, Arash Ahmadi and Majid Ahmadi

Abstract

In this chapter, we discuss the compute-in-memory memristive architectures and develop a 2M1M crossbar array which can be applied for both memory and logic applications. In the first section of this chapter, we briefly discuss compute-in-memory memristive architectural concepts and specifically investigate the current state off the art composite memristor-based switch cells. Also, we define their applications e.g. digital/analog logic, memory, etc. along with their drawbacks and implementation limitations. These composite cells can be designed to be adapted into different design needs can enhance the performance of the memristor crossbar array while preserving their advantages in terms of area and/or energy efficiency. In the second section of the chapter, we discuss a 2M1M memristor switch and its functionality which can be applied into memory crossbars and enables both memory and logic functions. In the next section of the chapter, we define logic implementation by using 2M1M cells and describe variety of in-memory digital logic 2M1M gates. In the next section of the chapter, 2M1M crossbar array performance to be utilized as memory platform is described and we conceived pure memristive 2M1M crossbar array maintains high density, energy efficiency and low read and write time in comparison with other state of art memory architectures. This chapter concluded that utilizing a composite memory cell based on non-volatile memristor devices allow a more efficient combination of processing and storage architectures (compute-in-memory) to overcome the memory wall problem and enhance the computational efficiency for beyond Von-Neumann computing platforms.

Keywords: compute-in-memory, crossbar, logic design, memory, memristor

1. Introduction

In general, memory devices are considered as one of the most important primitives in every computing system. Although, they play an undeniable role in conventional computers, which the processing units and memory are separate, it is generally believed that future computers, unlike von Neumann architectures, will have a compute-in-memory (CIM) structure. According to Moore's Law and fundamental VLSI limitations, CMOS technology is expected to face constraints and serious challenges at each technology node [1]. These challenges require solutions both short-term and long-term solving technical and strategic difficulties on Moore's Law way. Accordingly, researchers both in academia and industries are working hard on different available options and solutions from device up to architecture levels proposing incremental as well as revolutionary approaches. Regarding to this requirements, many efforts and initiatives have done by researches in order to keep on progress in the emerging memory technologies such as Ferroelectric Random Access Memory (FeRAM) [2], Magnetic Random Access Memory (MRAM) [3], and Resistive Random Access Memory (RRAM) [4], etc. Among all these technologies RRAM (generally referred as memristor) has received a lot of attention not only because of its favorable characteristics of low operating voltage, high speed, simple structure, and nano-scale but also with its logic implementation capabilities of the memristor devices. Memristor first in 1971, was proposed by Chua as a non-linear passive element [5], and then almost 37 years later, in 2008, was physically realized by the HP company, which was fabricated utilizing a Pt–TiO_2–Pt structure [6]. These nano-devices are based on a resistor with6variable resistance which can maintain resistance value upon bias removal that can be used as non-volatile memory cells. In addition to the conventional usage as memory cells, this device has also found variety of interesting applications such as machine learning platforms [7, 8], logic circuits design [9, 10], and neuromorphic systems [11]. Considering memristor as a non-volatile memory device makes it an interesting building block for large scale non-volatile memory systems. The memristor, or memory resistor, has been used in crossbar array architectures [12].

Due to the structural limitations (e.g. sneak path problems, interconnect resistance and etc.) of fully passive arrays (0T1M) various resistive switching memory based structures for the memory cells has been offered in literature such as 1T1M [13, 14], 4M1M [15], 1S1M [16–22], 1D1M [23], and 2T1M [24]. One challenging issue in crossbar array performance is sneak path current which can lead to negative effects on power consumption and limit the array size and other negative effects. Despite of amazing footprint size ($4F^2$) in fully passive crossbars, 1T1M arrays has been developed to reduce the impact of alternate currents with the cost of adding an access CMOS transistor in a single memory cell which significantly reduces the area efficiency of the array. These pseudo-crossbar structures mostly developed for digital memory arrays and they enable making large crossbars by adding more accessibility to each memory cell and avoiding the problem of voltage degradation over memory crossbar interconnects. 2T1M structure [24] is also presented and the auxiliary CMOS device is added to help for self-learning mechanism and these structure are used for spiking neural networks (SNNs). Also, a modified version of these cells are designed in 1T1M [13, 14] manner by getting benefit from the new type of transistor which has a smaller size and has the ability to change the sign of the charge carriers. Two terminal selectors such as non-linear switching elements and diodes are attracting a lot of attentions due to the scalability and small footprint sizes. Symmetric voltage–current characteristics for 1S1R structure in [17–22] avoid using these type of cells in logic applications. Also, for composite memory cells with diodes, Zener diode is utilized due to the low break down voltage which makes possible the rewriting over the memrisor device in each cell. Complementary resistive switch (CRS) with back to back memristor devices provide resiliency toward the sneakpath current by keeping one of the series device in high resistance which reduce the alternate current path in non-selected cells. Pure memristive composite memory array with 4M1M structure is proposed in [15], this structure provides a memristor switch to avoid sneakpath. In this method, at least one of the input devices is in low-resistance state (R_{on}) all the time which connects target cell to other cells in the row through a low resistive network. However, this structure suffers from parallel branches detour currents which considerably impact the power consumption and writing current.

Other attractive domain for composite memory structure is utilizing them for logic applications by collocating the computing within the memory in the same place. Several number of logic design and implementation research works have recently been proposed using memristor devices. Memristor Ratio Logic (MRL) is a CMOS-Memristor structure approach for combinational logic design [25]. In this method logical values are presented as node voltages, but it is a hybrid approach consisting both memristor and MOS transistors in the crossbar fabric. There are also other methods, such as MAGIC [26] and IMPLY [27] in which unlike MRL, memristance of memristors represent logical values. Each approach has positive and negative points regarding required number of memristor or MOS transistors or required time steps.

This chapter discuss 2M1M composite memory array and its application in both memory and logic. The proposed switch provides three modes namely, ON, OFF and No-Change, designed with three memristor. This structure not only can be used as AND, OR, NAND, and NOR logic gates with less computational steps compared to [27], but also the IMPLY logic can be implemented in crossbar array by this memory cell. The proposed cell is a pure memristor memory cell as 2M1M. The read and write operations are done by the same memristor circuits without need for additional circuitry within memory fabric. Thus, significantly reducing the number of required elements and simplifies the crossbar structure. The technique presented in the reading circuit does not need an isolated access to the memristor node which in turn reduces circuit wiring, and leads to a very simple structure with less complexity. Proposed structure provides an effective gating mechanism by which memory elements can be partially isolated from the access line during reading cycle which considerable reduces the sneak path currents. The remainder of this chapter is organized as follows: Section 2 introduces memristor-based switch circuit and its application and performance in the proposed memory cell. In Section 3 the proposed crossbar structure is discussed. Logic implementation and computational operations by 2M1M memory cell are presented in Section 4 and some explanation about sneak path are discussed.

2. A 2M1M memristor cell and its functionality

2.1 2M1M switch circuit

The 2M1M three state switch [9] which functions in ON, OFF, and NC are shown in **Figure 1**. As it can be seen, the proposed memory cell comprises of three memristor devices X_A, X_B and X_C. X_A and X_B devices are the access devices and they isolate the target device X_C which stores the information. There are three terminals A, B, and C in this structure in which A and B are considered as input terminals and V_a and V_b (as input voltages of '$-V$' or '$+V$') should be applied to terminals A and B respectively. Operation of the circuit, regarding V_a and V_b as input voltages and V_M as its output can be explained as follows. The input voltage ($+V$) for logic '1' and the input voltage ($-V$) for the logic '0' are applied to the X_A and X_B memristors ($|\perp V| > |V_{th}|$). The voltage V_M on common node of memristors represents output of the circuit while memristor X_C maintains this value in form of memristance. The truth table of this circuit is shown in **Table 1**.

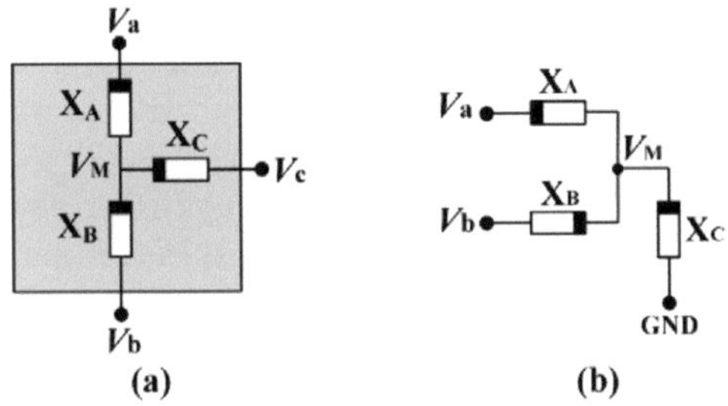

Figure 1.
Schematic of the proposed memory cell. (a) General circuit of memory cell. (b) Configuration for switch circuit mode. Figure reprinted by [9].

Case	V_a	V_b	V_M	R_c	Switch State
1	$-V$	$-V$	$-V$	R_{off}	OFF
2	$-V$	$+V$	0	R_{ini}	NC
3	$+V$	$-V$	0	R_{ini}	NC
4	$+V$	$+V$	$+V$	R_{on}	ON

Table 1.
Truth table for the proposed switch circuit.

$$\frac{(V_M - V_a)}{R_a} + \frac{(V_M - V_b)}{R_b} + \frac{V_M}{R_c} = 0$$
$$\rightarrow V_M\left(\frac{1}{R_a} + \frac{1}{R_b} + \frac{1}{R_c}\right) = \frac{V_a}{R_a} + \frac{V_b}{R_b} \tag{1}$$

and providing R_a, $R_b << R_c$ we can approximate V_M as:

$$V_M = \frac{R_b}{R_a + R_b} V_a + \frac{R_a}{R_a + R_b} V_b \tag{2}$$

When both inputs are '0' ($-V$) according to the polarity of memristors, since a negative voltage is applied across memristor X_A and a positive voltage across memristor X_B, so their memristance, regardless of their initial states, will change to Roff and Ron, respectively. Therefore, according to Kirchhoff's law and also considering the initial state of the memristor X_C, as $R_C >> R_A$, R_B (memristance of X_A and X_B), the voltage in common node of memristors is: $V_M = -V$. This will set memristance of X_C to R_{off}, which is logical zero:

$$V_a = V_b = -V$$
$$V_M = \frac{R_a + R_b}{R_a + R_b} \cdot (-V) = (-V) \approx Logical\ 0 \tag{3}$$

For logic 1, according to the fourth row of the **Table 1**, when both inputs are in the same value of $+V$, similarly, based on the polarity and direction of memristors, the memristor X_A is set to R_{on} and memristor X_B becomes R_{off}. Therefore, the voltage on the output node (M) is approximately $+V$ which will change the memristance of the output memristor, X_C, to R_{on} representing logical one:

$$V_a = V_b = V$$
$$V_M = \frac{R_a + R_b}{R_a + R_b} \cdot V = V \approx Logical1 \tag{4}$$

Otherwise, if the value of input voltages is different as (+*V*) and (−*V*), both input memristors have the same value of either R_{on} or R_{off} according to the applied voltage and their polarity. This results in a zero voltage on common node. Since $V_c = 0$, in this case memristance of X_C does not change:

$$V_a = -V_b = V$$
$$V_M = \frac{R_{on}}{R_{on} + R_{on}} \cdot V + \frac{R_{on}}{R_{on} + R_{on}} \cdot (-V) = 0 \tag{5}$$

or similarly:

$$V_a = -V_b = (-V)$$
$$V_M = \frac{R_{off}}{R_{off} + R_{off}} \cdot (-V) + \frac{R_{off}}{R_{off} + R_{off}} \cdot V = 0 \tag{6}$$

This state is called a NO-Change state. To have a timing analysis of switches operation, according to [15]:

$$\frac{dR(t)}{dt} = -k \cdot i(t) = -k \cdot \frac{V}{R(t)} \tag{7}$$

$$k = \mu.\Delta R \cdot R_{on}/D^2 \tag{8}$$

$$\Delta R = R_{off} - R_{on} \tag{9}$$

Because in the fourth combination of the truth table of memristor based switch (**Table 1**), memristor X_A is parallel with memristor X_B then $V_a = V_b$. Therefore:

$$V_a = \frac{R(t) \cdot dR(t)}{-k \cdot d(t)} \tag{10}$$

By integrating (8) and also assuming that $\phi_0 = 0$, $\phi(t)$ is given by

$$\int Va = \int \frac{R(t) \cdot dR(t)}{-k \cdot d(t)} = \int \frac{d\phi}{dt} \tag{11}$$

$$R_a{}^2(t) - R_{ai}{}^2 = -2k_a\phi(t) \tag{12}$$

$$\phi(t) = \frac{(R_a{}^2(t) - R_{ai}{}^2)}{-2k_a} \tag{13}$$

and also by supposing the initial state of memristor X_A is R_{off} and its final state is R_{on}, the required flux across the memristor X_A is

$$\phi(t) = \frac{(R_{on}{}^2 - R_{off}{}^2)}{-2k_a} \tag{14}$$

Thus, the required time for change state of memristor X_A and X_B is given by:

$$\Delta\phi_a = V_a T_1 \tag{15}$$

$$T_1 = \frac{\left(R_{on} + R_{off}\right) \cdot D^2}{-2V_a \mu_v R_{on}} \tag{16}$$
$$T_1 = T_2$$

In memristor X_C the required time for the change of state is:

$$T_3 = \frac{\left(R'_{on} + R'_{off}\right) \cdot D^2}{-2V_C \mu'_v R'_{on}} \tag{17}$$

Therefore, the total time to change the cell state, T_t, is given by:

$$T_t = T_1 + T_3$$
$$T_t = \frac{\left({R_{off}}^2 - {R_{on}}^2\right)}{-2k_a V_a} + \frac{\left(R'_{on} + R'_{off}\right) \cdot D^2}{-2V_C \mu'_v R'_{on}} \tag{18}$$

2.1.1 Simulation result for 2M1M switch circuit

This is in general agreement with the simulation results as presented in follow, Despite several memristor SPICE models which are presented in [28, 29], the simulation results are performed using Biolek model presented in [30]. This model is selected due to the fact that, it can be utilized in mathematical analysis for power and delay estimation besides its validity to characterize the memristor switching behavior. PSPICE software has been utilized to perform the simulations. The simulations are carried out by using the parameters in **Table 2**, and for a fair comparison, these parameters are similar with [15] to evaluate functionality of the design.

Different combinations of inputs which are applied to the switch are shown in **Figure 2**. As it can be seen the simulations results are in agreement with the truth table of **Table 1**. Here the voltage is applied and output logics is represented by memristance of the X_C. Delay or settling times for this switch is defined by the time which X_C memristance reaches to its final value. According to the simulation results, this time is 1.11 ns which is also in agreement with theoretical calculations.

2.2 Write and read operations

For the write operation, the memory cell should work based on the first and forth rows of **Table 1**, respectively for writing '0' and '1', as descried in details in subsection 2.1. For read operation, unlike previous works, these cells do not need any additional wiring or complicated sense circuitry. This is because in this circuit

Parameters	Value	Parameters	Value
D(nm)	1	V_{th}(V)	0.11
V_w(V)	0.9	V_R(V)	0.1
R_{on}(Ω)	100	R'_{on}(kΩ)	10
R_{off}(Ω)	900	R'_{off}(Ω)	1900
μ(m^{2}V^{-1} s^{-1})	1×10^{-6}	μ'(m^{2}V^{-1} s^{-1})	1×10^{-7}

Table 2.
Simulation parameters for 2M1M memory architecture.

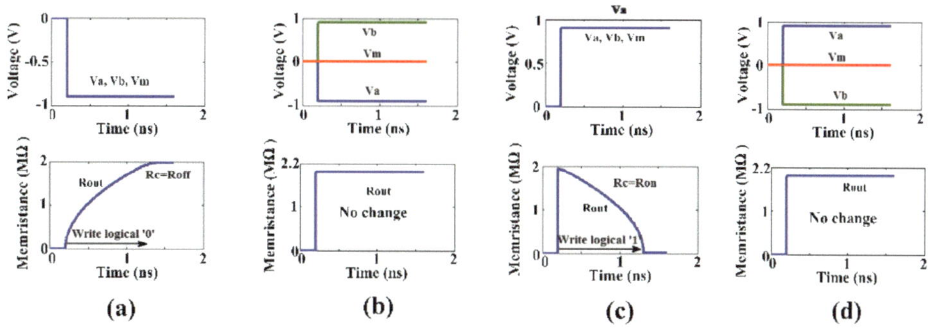

Figure 2.
Memristor switch circuit simulation results for different cases. In each subfigure, upper figure displays the inputs and common node voltages while the below figure displays the output device resistance state. (a) a = 0, B = 0. *(b)* a = 0, B = 1. *(c)* a = 1, B = 0. *(d)* a = 1, B = 1. *Figure reprinted by [9].*

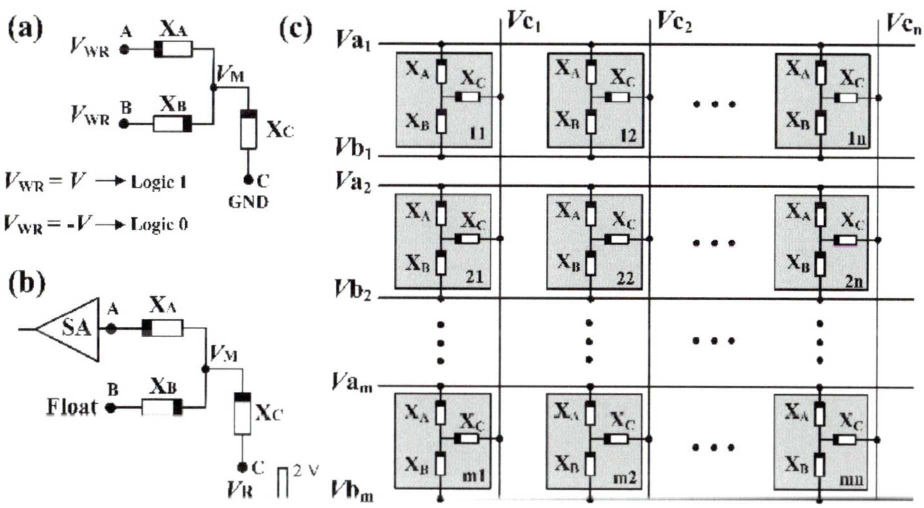

Figure 3.
(a) Write and (b) read circuit configurations of the proposed memory cell. (c) 2M1M Memristor-based crossbar architecture. Figure reprinted by [9].

memristor change inertia, as shown in **Figure 3**, is exploited. In general, the READ operation is done by floating node B and applying READ signal V_R to node C. Then, a sense amplifier (SA) is sensed the current from node A of the memory cell. In other words, in this technique a read signal is applied to the memristor which does not change the memristance, either because of its high frequency ($f > f_{th}$) or/and because of its low voltage ($V < V_{th}$).

In read process method, a pulse *VR* with appropriate amplitude and small duration is applied to the circuit as "read signal". A sense amplifier then measures value of the propagated signal on port A of the switch. Width and amplitude of this pulse (or spike) should be chosen in a way to do not affect memristors' state during read process. As mentioned before in high frequencies memristor operates like a pure resistive element. If we connect A and B ports of the proposed circuit to the ground, and apply to the other end of the memristor X_C (port C), a read spike with amplitude voltage of 2 V, as shown in **Figure 3b**, the output voltage can be read at node VM as:

$$V_M = \frac{(R_{on} \| R_{off})}{(R_{on} \| R_{off}) + R_c} \cdot 2V \cong \frac{R_{on}}{R_{on} + R_c} \cdot 2V \tag{19}$$

which in terms of logic vales can be described as:

$$\begin{aligned} if: R_c = R_{on} \rightarrow V_M = V = \text{logical} \quad 1 \\ if: R_c = R_{off} \rightarrow V_M \cong 0 = \text{logical} \quad 0 \end{aligned} \tag{20}$$

With this technique we can increase reading speed and reduce power consumption. In addition, if we read the output value from port A instead of node M, this read method does not require any additional wiring to access node M. This considerably reduces fabrication and wiring complexity of the proposed crossbar structure.

3. 2M1M Memristor crossbar architecture

In crossbar architecture the 2M1M memory cell can be used effectively as shown in **Figure 3c**. While, there is no need CMOS transistors for each cell within cross bar fabric in this architecture. As it can be seen in **Figure 3c**, the similar nodes of memory cells in the crossbar structure are connected to each other in the horizontal rows nodes A_i and B_i are of the cells are connected to each other separately while in vertical columns modes C_i are connected to each other. The desired reading or writing operation are performed by applying appropriate voltages in suitable rows and columns to activate a cell and disable others.

3.1 Write operation in the crossbar

For write operation in the crossbar architecture, like a single memory cell, appropriate input values need to be applied to the memory cell based on **Table 1**. This means that, both applied voltages V_a and V_b should be similar, either amount of '$+V$' or '$-V$' to write logical '1' or to write logical '0' respectively. Otherwise, the other states in truth table, the switch is in the No-Change state. This scheme is easily applicable to the crossbar structure in the same way as a single cell using connected cell port (a_i, b_i and c_i). It should be considered that when read or write signal are applied, cells should be completely isolated from the target cell. When writing in a cell (or a number of associated cells as a word), the other cells should have maintained their saved vales. For more explanation for write operation, as an example. Considering cell 22 as a target cell to write '0' ('1') in **Figure 4**. In this case, the same voltages $-V$ (similarly $+V$ for '1') should be applied to both memristors X_A and X_B and node C is connected to GND. In this situation, in terms of applied voltages

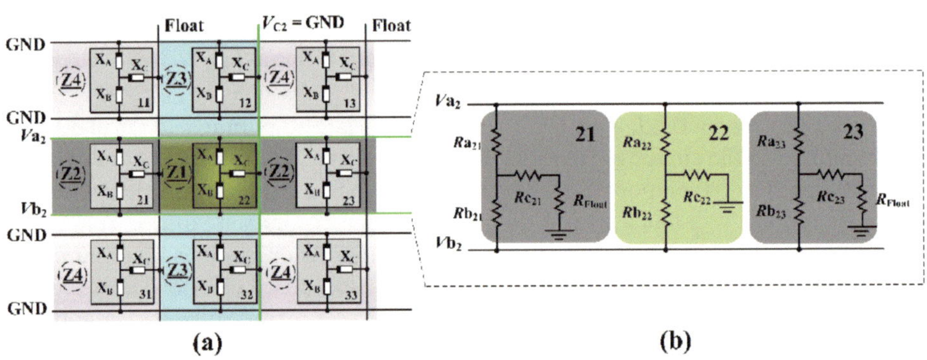

Figure 4.
(a) Write configuration in 3 × 3 2M1M crossbar memory and the hazardous zones are displayed. Z1 zone contains target cell 22 and highlighted by green. (b) Resistive equivalent circuit of zone Z_2 in the proposed 3 × 3 2M1M crossbar memory. The equivalent resistor circuit of the target cell 22 is highlighted by green. Figure reprinted by [9].

combination, four different zones are recognizable in the crossbar structure, as shown in **Figure 4**. As it can be seen only the target cell is located in the first area, Z_1. The second area is Z_2, where all cells have the same voltages as target cell on their a_i and b_i nodes. In the third area, Z_3, cells have the voltage on their node C, which is the same as the target cell 22. In the fourth zone, Z_4, there is no input in common with cell 22. In the Z_1 to write '0' ('1') into the cell 22 voltage $-V$ ($+V$) is applied to rows V_{b2} and V_{a2} where the column V_{c2} is connected to GND.

As can be seen in the **Figure 4**, Z_2 is the hazardous zone because in this area same voltages as the target cell ($-V$ or $+V$) are applied to the V_{a2} and V_{b2} ports of the cells. It can cause an unwanted writing and changing the state of the memristors that are not supposed to change. To deal with this issue, C nodes of the neighbor cells in zone Z_2 are floated or in practice connected to a high impedance open circuit (the columns V_{c1} and V_{c3} in **Figure 4**). Since the columns c_1 and c_2 are floated, we consider a resistance R_{Float} for each of these columns and this resistance is connected to the non-bar side of the X_C device in each of 21 and 23 2M1M memory cells. The bar side of X_C device is connected to the common node between X_A and X_B devices. Then, the equivalent resistor-based circuit for cell 21 has two serially connected resistors R_{a21} and R_{b21} which are connected to rows a_2 and b_2 and they are memristances of X_A and X_B devices in cell 21, respectively. The common node of X_A and X_B is connected to a R_{c21} resistor which is a memristance of Xc device in cell 21. Therefore, resistance of the float column R_{Float} will be in series with R_{c21} resistor (this is true for cell 23). If the float resistance terminals were connected to both terminals of X_C device, then we could consider R_{Float} was parallel with R_{c21} while here only the non-bar side of X_C device is connected to the floated column c_1.

Resistor equivalent circuit of this zone is depicted in **Figure 4b**. This floated port connection reduces current through X_C memristor of the unselected cells (≈ 0) which keeps the stored values of the cells untouched. The rest of the rows and columns in the cross-bar structure are connected to ground. Thus, points A, B and C of cells that are in zones Z_3 and Z_4 are either floating or connected to GND. the logical state of these cells therefore do not change during write operation. Although, maybe one of the two memristor X_A and X_B is sufficient to perform write operations and by help one of them could to done correctly write operation but as mentioned, this structure is designed to be based on a three-state switch ON, OFF and NO Change. The second case is used to high impedance operation for memristors without changing of output memristor in practice reading that through this can reduce the sneak paths current. In addition, one of the applications of these cells has been mentioned is the implementation of logic circuits which is explained in the Section 4. Please note that in this structure cells in the same column are almost independent and can be written or read simultaneously. This makes it possible to have a parallel read/write process on these cells for higher rate memory access operations or combine a number of them forming data "word" rather than collections of single bits.

In the write operation as can be seen in **Figure 4a**, the memory cells 21 and 23 are in zone 2 and they are the neighboring cells of the target cell for write operation. The equivalent resistor-based circuit of these cells are displayed in **Figure 4b**. Write operation of the memory cell in 3×3 2M1M crossbar array is simulated in **Figure 5a** and **b**. This memory cell is functioning even by having a time difference between the applied voltage input V_a and V_b. To test the proposed memory cell for this special case, two asynchronous input voltages are applied to the memory cell and the simulation results prove its functionality (**Figure 5c**).

3.2 Read operation in the crossbar

Regarding read operation, there are four zones in the crossbar as described in previous section. During read operation, as shown in **Figure 6a**, by applying voltage

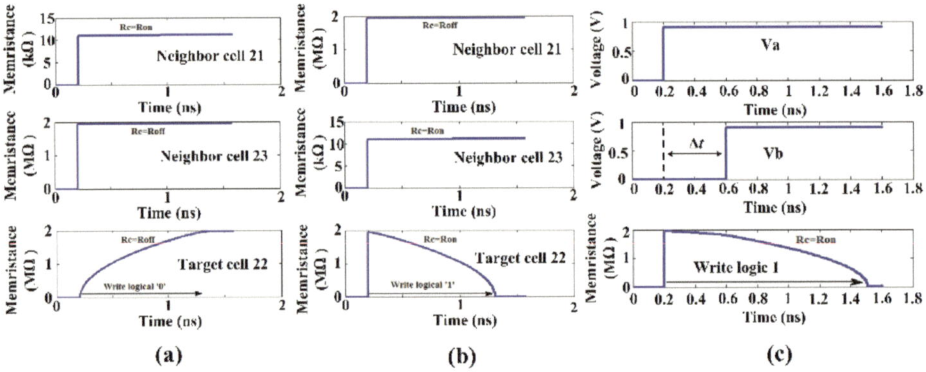

Figure 5.
Writing in target cell with different neighbor cell's stored value. (a) Write 1 in target cell. (b) Write 0 in target cell. (c) Write operation when two asynchronous input voltages are applied to the target cell with the time difference. Figure reprinted by [9].

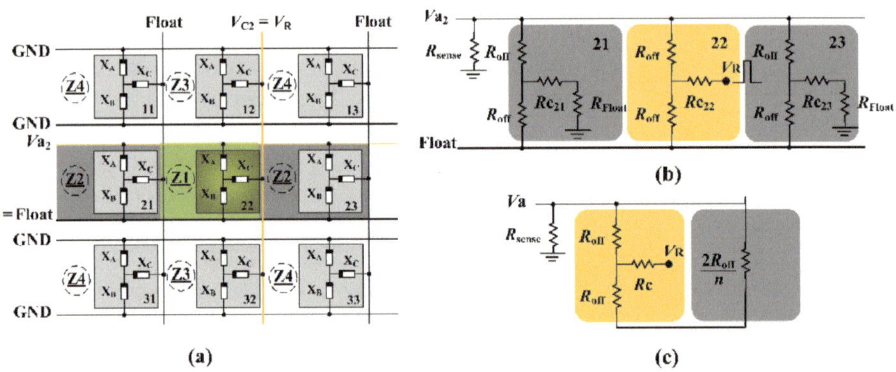

Figure 6.
Read configuration in 3×3 2M1M crossbar memory and the hazardous zones are displayed. (a) Z1 which contains target cell 22 is highlighted by green. Equivalent resistive circuit for hazardous zone for read operation. Target cell for read operation is highlighted by orange. (b) Resistive circuit for hazardous zone in 3×3 2M1M crossbar memory. (c) Equivalent resistive circuit by considering n neighbor cells. Figure reprinted by [9].

V_R to node C and stored bit in X_C can be read as a voltage from node A. Suppose that we want to read from cell 22. The read operation must be performed in 2 stages; in the first stage, memristors X_A, X_B of the Z_2 memory cells are changed to high impedance (R_{off}) to partially isolate neighbor cells in this zone from applied read spike which can be done by applying voltages $-V$ and $+V$, according to truth table of **Table 1**, to lines V_a and V_b of the cells in the zone respectively. At second stage the read signal is applied to port C of the target cell and voltage of port A of the cell is read. This stage must be performed by floating row V_{b2}, applying voltage V_C to column V_{c2}, reading (measuring) the voltage on node A using a sense amplifier. The important point at this stage is considering appropriate signal as read signal. It is very important that applied read pulse be strong enough to induce a readable voltage at A line of the row. And also this signal should not affect memristance values of the memristors in the target cell or the neighbors. Here a spike shaped narrow pulse is used as read signal (V_C).

Another consideration which is so important in this crossbar architecture is effect of neighbor cells in the output readout value. In this case the circuit can be assumed as a resistive network and areas involved in this operation are Z_1, Z_2 that can be seen in **Figure 6**. Sneak path current is considered as one of the most important issues in memristor crossbar memories. Here, X_A and X_B of the neighbor cells to "gate" effect of the X_C memristance of the neighbors from read signal are used. By changing memristance of X_A and X_B memristors of the neighbors to R_{off}, as shown in **Figure 6b**, the target cell will be in parallel connection with its neighbors which are gated form a_i line by $2R_{off}$ memristance. Since all cells in these areas, except cell 22, have a floating (R_{float}) resistance connected to the node C, each neighbor cell can be considered as a 2Roff resistors in parallel with cell 22. The value of these parallel resistances is equivalent to $(2/n) \times R_{off}$ (n = total number of columns per row). Accordingly, equivalent resistance of the neighbor cells from ai line is almost independent from their X_C memristance, which represents stored value in the cell. This technique considerably reduces sneak path effect and its negative effect on cells' readout process. Using equivalent circuit of **Figure 6c**, the readout voltage and equivalent neighbor cells resistance can be calculated as:

$$V_a = \frac{R_{sense}}{((1+\frac{2}{n})R_{off})\|R_{off}) + R_c + R_{sense}}.V_c \tag{21}$$

$$\left(1+\frac{2}{n}\right)R_{off})\|R_{off}) \cong \frac{R_{off}}{2} \tag{22}$$

by selecting of R_{sense} = R_{off},

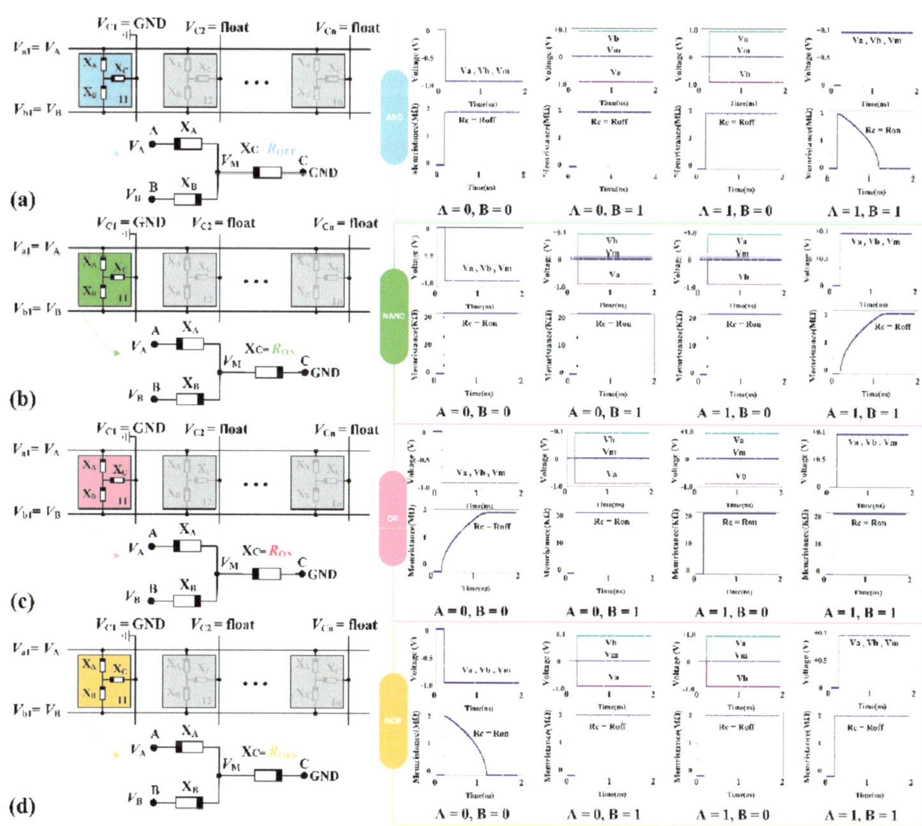

Figure 7.
2M1M array logic schematics and simulation results for AND, NAND, OR, and NOR. (a) AND logic gate for 2M1M switch. (b) NAND logic gate for 2M1M switch. (c) OR logic gate for 2M1M switch. (d) NOR logic gate for 2M1M switch.

$$
\begin{aligned}
&V_a = \frac{R_{off}}{\frac{R_{off}}{2} + R_c + R_{off}} \cdot V_c \\
&if : R_c = R_{on} \Rightarrow V_a = \frac{R_{off}}{\frac{R_{off}}{2} + R_{on} + R_{off}} \cdot V_c \approx \text{Logical } '1' \\
&if : R_c = R_{off} \Rightarrow V_a = \frac{R_{off}}{\frac{R_{off}}{2} + R_{off} + R_{off}} \cdot V_c \approx \text{Logical } '0'
\end{aligned} \tag{23}
$$

The second voltage V_a from node A is higher or lower voltage according to X_c which is low or high resistance state (R_{on} or R_{off}). **Figure 7**, presents simulation results for a read operation in the crossbar structure. As it is discussed, to read the stored value of a cell we have to apply a spike like pulse to the C node of the cell and read the voltage from line a_i, where bi line of the row and C node of the other cells in the same row are float. When R_c is R_{off} the voltage in node A is a low voltage that is equivalent to logic zero and vice versa, when R_c has the value of R_{on}, the voltage in node A has a higher voltage which represents to a logic one. **Figure 8**, presents two different cases. **Figure 8a**, shows reading '0' from a cell, when the neighbor stored '1' **Figure 8b**, shows reading '1' from a cell, when the neighbor stored '1'. In both cases target cell has been readout correctly. According to the simulation results

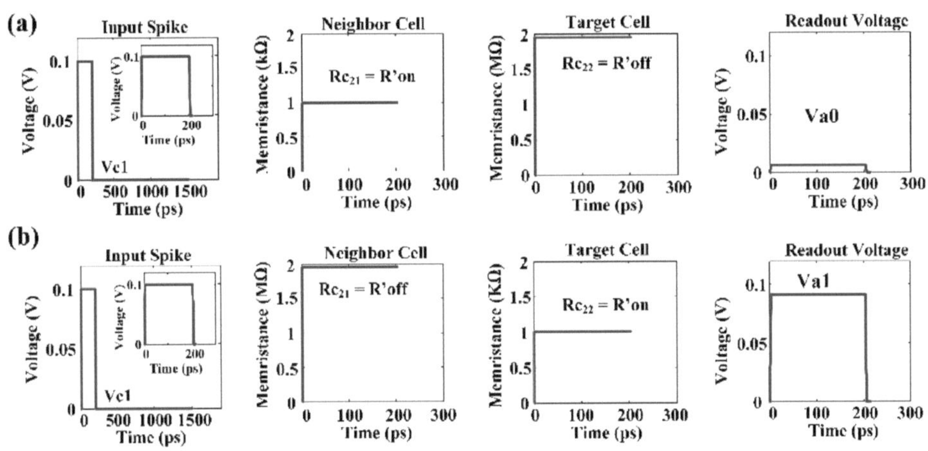

Figure 8.
The simulations of memory cell in crossbar array for; (a) read of logic 0 from target cell and other cell, (b) read logic 1 from target cell and other. Figure reprinted by [9].

	Memory [23]	Memory [24]	4M1M [15]	2M1M
Read time	1.2 ns	1.095 ns	0.25 ns	20 ps
Write voltage	1.0 V	0.9 V	0.9 V	0.9 V
Read voltage	±1.0 V	0.9 V	0.1 V	0.1 V
Number of consecutive read	—	130	$\gg 10^5$	$\gg 10^5$
Number of consecutive read by 10% noise	20	—	$\gg 10^5$	$\gg 10^5$

Table 3.
Comparison of read operation with previous works.

and the formula presented in the [16], reading margin in this work is equal to the amount 0.7 V which can be a reasonable amount.

$$\mathrm{RM} = \frac{\Delta Vout}{\Delta Vread} = \frac{Vout(LRS) - Vout(HRS)}{V_{WS}} \tag{24}$$

where V_{WS} is the read voltage applied. Simulation results are compared with previous works in **Table 3**. As it can be seen in this table, the proposed method is considerably better than [31, 32], and is similar to [15].

4. Logic implementation and computational operations by 2M1M memory cell

Composite memory cells can be applied to implement digital logics. In addition to its memory application, the proposed memory cell is capable of implementing logic which makes it capable for in-memory computing applications. Here, in this section we are assessing the logic implementation of the proposed architecture with 2M1M cells.

4.1 Logic gates with 2M1M switch

From switching point of view, this circuit is a three state switch as 'ON', 'OFF' and 'No-Change'. Interestingly, this switch can also be used as logic gates. By setting the initial memristance value of the output memristor to R_{on} or R_{off}, final memristance state of memristor X_C, respectively, AND or OR logic gate operations are developed. Further, by changing the polarity of the output memristor (X_C) one can make NAND and NOR gates in a similar way. Therefore, the 2M1M array can develop two different logic schemes based on the polarity of memristor X_C. First, include AND and OR gates and by changing the polarity of X_C the array can develop NAND and NOR gates. The logic is based on the resistance of device and not the voltage. This will make this logic to enable in-memory compute logic family as the data will store within the memory array after finishing the operation.

The input voltage pulses with amplitude $+V$ and $-V$ are applied as logic 1 and 0 into the rows a_1 and b_1. Other unselected rows will be floated and the column c_1 is grounded to shape a 2M1M cell 11 as a logic gate. Other unselected columns need to be floated to inactive the rest of the 2M1M cells in the corresponding row. As an example, the AND gate can be implemented by a 2M1M switch over the 2M1M array by applying the appropriate voltages. This gate is comprised of two access devices X_A and X_B which are connected in parallel with different polarities to node *M*. The output device X_C is connected between node M and bit-line of the array by a positive polarity. The input voltages should be applied to a_1 and b_1 lines as V_A and

A	B	AND		OR		NAND		NOR	
		Vo	Ro	Vo	Ro	Vo	Ro	Vo	Ro
$-V$	$-V$	0	R_{OFF}	0	R_{OFF}	1	R_{ON}	1	R_{ON}
$-V$	$+V$	0	$R_{IN} = R_{OFF}$	0	$R_{IN} = R_{ON}$	0	$R_{IN} = R_{ON}$	0	$R_{IN} = R_{OFF}$
$+V$	$-V$	0	$R_{IN} = R_{OFF}$	0	$R_{IN} = R_{ON}$	0	$R_{IN} = R_{ON}$	0	$R_{IN} = R_{OFF}$
$+V$	$+V$	1	R_{ON}	1	R_{ON}	0	R_{OFF}	0	R_{OFF}

Table 4.
Truth table of the proposed memristor logic gates.

		IMPLY [27]	4M1M [15]	2M1M
Operation step	AND	4	1	1
	OR	3	1	1
	NAND	3	2	1
	NOR	4	2	1
Required memristors	AND	4	3	3
	OR	3	3	3
	NAND	3	4	3
	NOR	3	4	3

Table 5.
Comparison of proposed 2M1M logic gates with [15, 27].

V_B. Also, $R_C >> R_A$, R_B and the resistance of R_C will specify the output of the logic. The logic can be described for different input combinations by considering the Eqs. (1)–(5).

The truth table of different 2M1M logic gates has been in presented in **Table 4**, by showing different input combination voltages, output voltage and resistance state of the output device. Different 2M1M logic cells, their implementations on memristor crossbar array and the simulation results corresponding to each AND, NAND, OR, and NOR logic gates by using 2M1M cells for different input combinations have been displayed in **Figure 7**. In **Table 5**, the proposed 2M1M logic gates have been compared in terms of number of with IMPLY logic [27] and 4M1M [15]. It has been shown that the proposed logic requires only one computational step to implement in-memory logic for AND, NAND, OR, and NOR gates. Also, the number of required devices to implement all of these logic gates are 3 devices included in a 2M1M cell structure.

Sneak path current is considered as one of the important challenges against practical application of memristor crossbars. During reading operation the sneak path currents through neighbor cells can affect readout value of the target cell. To eliminate or reduce the sneak path in the crossbar array several methods have been proposed by researchers. In general, proposed methods can be divided into two categories. In the first approach [33–36], researchers focus on device level structure of the memristor or read process in the crossbar to make it more resilient against this effect. Among these methods is the way provided in [33] in which read operation is done by an algorithm in several stages. This method improves the sneak path problem but increases read time and require additional circuit to realize the read algorithm stages. Another approach relies on memristive devices with inherent nonlinear structure such as [34, 35] in which a three-terminal memristor device is proposed to solve this problem. In another approach as presented in [36] to eliminate sneak path currents separate columns are considered for each element in the crossbar architecture. That increases cell area and therefore reduces the memory density. In the second approach, to solve the problem of sneak path currents, it is suggested to add additional switches to each memory cell in the crossbar architecture to separate reading path of the target cell from the other unwanted paths. There are several suggestions in this approach, but the most popular structure is 1T1M (one transistor for one memristor) [14]. This structure uses a transistor to separate each cell from other cells during read operation. In this way, added transistor is the gating element of the cell. This method has problems due to the scalability considerations of the CMOS-memristor structure [13]. In [23] diodes, instead of transistors, have been suggested to reduce sneak path. There are difficulties with this approach as well due to diode behavior. In another approach [28], back-to-back memristors are proposed to overcome the problem of sneak path current in which always one of the memristors is in R_{off} state and the other one is R_{on}. In this way, equivalent memristance is always greater than R_{off} which can reduce the sneak path effect. In [15] a memristor based switch is suggested to solve the problem of sneak path. In this method at least one of the input memristors is always in state R_{on}, which connects target cell to other cells in the row through a low resistive network. In this structure, as shown in **Figure 9**, during write process, there is a detour current path through M_S-M_P and M_T-M_P, in all parallel branches in the crossbar structure; witch can considerably increase the writing current and power consumption.

In this study, effect of sneak path can be easily reduced using proposed gating mechanism created by X_A and X_B memristors in the cell. By changing state of these memristors to R_{off}, memristor X_C, which keeps the saved value ('0' or '1') of the memory cell, can be isolated from rest of the network. As discussed before, in the second and third rows of the truth table switch goes to No-Change state and X_C keeps its sate untouched, where both X_A and X_B memristors become either R_{off} or R_{on}. Therefore, if in the crossbar array structure, we apply $-V$ and $+V$ to a_i and b_i lines of the row respectively, memristors X_A and X_B of all the cells in the row will to R_{off} state, which is a high impedance, without any change in their X_C memristance. So unlike cells provided in [15] there is no resistance of R_{on} between the selected node and the other nodes of the circuit. In fact, high impedance of the X_A and X_B memristors isolate X_C of all the cells in the crossbar from each other.

With this approach equivalent circuit of the neighboring cells in a row is as shown in **Figure 9**. Interestingly, target cell (first cell from left) sees an equivalent resistor of the network which is almost independent from stored values (in terms of R_{on} or R_{off}) in other cells. This means if $R_{\text{float}} >> R_{\text{off}}$ then effect of R_c state is negligible on I_{read} current. Simulation results are presented in **Table 6**. As it can be seen this method is far better than [15, 24]. In comparison with [31] sneak path current in this work is higher but please note that in [31] there are two transistors for each memory cell but our cell is transistor-less.

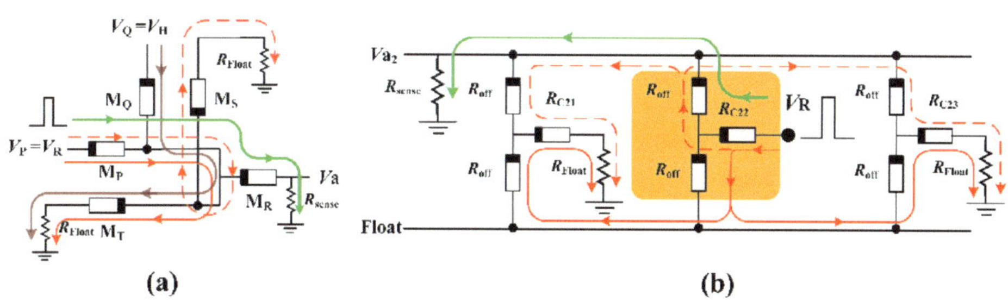

Figure 9.
Sneak path current in 4M1M cell [15] and the proposed 2M1M crossbar memory in a read operation. (a) 4M1M [15] sneak path currents in read operation. (b) Sneak path current in 2M1M crossbar during read operation. Figure reprinted by [9].

	1T1M [13]	2T1M [24]	4M1M [15]	2M1M
Sneak current	0.25–33.29 μA	5.0 pA	0.24–1.77 μA	9 nA
State change	0.261	0	0	0

Table 6.
Comparison of sneak current effect of the proposed architecture with other architectures.

	SRAM [37]	Memory [38]	Memory [16]	4M1M [15]	2M1M
Density (Gbt/cm^2)	0.338	1.6	—	50	80
Energy (fj/bit)	28.4	—	0.011	2.5×10^{-4}	23.2×10^{-9}

Table 7.
Comparisons of density and energy consumption with previous works.

By providing the structure and strategies for array-based 1S1R [16–22], many of the structures have offered while having high density $4F^2$, very small sneak paths current, very low power consumption and high read margin that is very promising. Compared with 2M1M structure, can be said that 1S1R based structures has been created in series connection memory element and selector in terms of manufacturing technology because are of the two different types perhaps compared with 2M1M structure which is a memristor uniform structure be more complexity. And in the 2M1M structure used of memristor, that is a memory and a computing element. The aim is to implementation the logic and computing capabilities for future applications of this structure in memory which can help to achieve a beyond classical von Neumann architecture. It hopes that by development and progression of 2M1M, the valuable feature in 1S1R structure is achieved for a higher density and removes sneak paths. Approximated device density and power consumption of the proposed architecture is compared with previous works in **Table 7**. As it is attainable form this table, due to lower number of memristors per memory cell, proposed architecture offers higher density compared with previous works. In terms of power consumption, since authors did not find a clear explanation regarding details of previous studies for their power calculations, power consumption of the cells in various operations are presented and compared in details.

5. Conclusions

In summary, this chapter discusses the resistive switching based composite memory cells and offers a solution toward the limitations within the current state-of-art 0T1R fully passive arrays and 1T1R active arrays to implement more efficient compute-in-memory structure for future beyond von-Neumann computing architectures. The first section of this chapter briefly review different resistive switching based composite memory arrays and discusses their advantages and limitations toward compute-in-memory applications and implementations. The next section, define a 2M1M memory array cell and analyzes its switching characteristics and the write and read operation principles within the crossbar structure. The final section of the chapter discusses the logic application with 2M1M switch and its capability to implement AND, NAND, OR, and NOR logic gates within 2M1M memory array structure and its compute-in-memory feature. Also, this section discusses the problem of sneakpath within the composite memory arrays and 2M1M array structure. We hope this chapter provide a good basis toward development of resistive switching based composite memory array platforms and providing a good insight over 2M1M structural benefits for compute-in-memory applications.

Author details

Mehri Teimoory[1], Amirali Amirsoleimani[2], Arash Ahmadi[3] and Majid Ahmadi[4*]

1 Department of Electrical Engineering, Razi University, Kermanshah, Iran

2 Department of Electrical Engineering and Computer Science, Lassonde School of Engineering, York University, Toronto, ON, Canada

3 Department of Electronics, Carleton University, Ottawa, ON, Canada

4 Department of Electrical and Computer Engineering, University of Windsor, Windsor, ON, Canada

*Address all correspondence to: ahmadi@uwindsor.ca

References

[1] International Technology Roadmap for Semiconductors. Available online at: http://www.itrs.net (Nov. 2015).

[2] Ducharme S, Reece TJ, Othon CM, Rannow RK. Ferroelectric polymer Langmuir-Blodgett films for nonvolatile memory applications. IEEE Transactions on Device and Materials Reliability. 2005 Dec;**5**(4):720-735

[3] Li H, Chen Y. An overview of non-volatile memory technology and the implication for tools and architectures. In2009 design, Automation & Test in Europe Conference & Exhibition 2009 Apr 20 (pp. 731-736). IEEE.

[4] Chua L. Resistance switching memories are memristors. Appl. Phys. 2011; A, 102(4):765–783.

[5] Chua L. Memristor-the missing circuit element. IEEE Transactions on Circuit Theory. 1971 Sep;**18**(5):507-519

[6] Strukov DB, Snider GS, Stewart DR, Williams RS. The missing memristor found. Nature. 2008 May;**453**(7191):80-83

[7] Amirsoleimani A, Alibart F, Yon V, Xu J, Pazhouhandeh MR, Ecoffey S, et al. In-memory vector-matrix multiplication in monolithic complementary metal–oxide–semiconductor-memristor integrated circuits: Design choices, challenges, and perspectives. Advanced Intelligent Systems. 2020 Nov;**2**(11): 2000115

[8] Cai F, Correll JM, Lee SH, Lim Y, Bothra V, Zhang Z, et al. A fully integrated reprogrammable memristor–CMOS system for efficient multiply–accumulate operations. Nature Electronics. 2019 Jul;**2**(7):290-299

[9] Teimoori M, Amirsoleimani A, Ahmadi A, Ahmadi M. A 2M1M crossbar architecture: Memory. IEEE Transactions on Very Large Scale Integration (VLSI) Systems. 2018 Feb 14;**26**(12):2608-2618

[10] Amirsoleimani A, Ahmadi M, Ahmadi A. Logic design on mirrored memristive crossbars. IEEE Transactions on Circuits and Systems II: Express Briefs. 2017 Jul 19;**65**(11): 1688-1692

[11] Rahimi Azghadi M, Chen YC, Eshraghian JK, Chen J, Lin CY, Amirsoleimani A, et al. Complementary metal-oxide semiconductor and memristive hardware for neuromorphic computing. Advanced Intelligent Systems. 2020 May;**2**(5):1900189

[12] Williams RS. How we found the missing memristor. IEEE Spectrum, vol. 45. No. 2008;12:28-35.

[13] Kim S, Jeong HY, Kim SK, Choi SY, Lee KJ. Flexible memristive memory array on plastic substrates. Nano Letters. 2011 Dec 14;**11**(12):5438-5442

[14] Zangeneh M, Joshi A. Design and optimization of nonvolatile multibit 1T1R resistive RAM. IEEE Transactions on Very Large Scale Integration (VLSI) Systems. 2013 Sep 10;**22**(8):1815-1828

[15] Zhang Y, Shen Y, Wang X, Cao L. A novel design for memristor-based logic switch and crossbar circuits. IEEE Transactions on Circuits and Systems I: Regular Papers. 2015 May;**62**(5): 1402-1411

[16] Zhou J, Kim KH, Lu W. Crossbar RRAM arrays: Selector device requirements during read operation. IEEE Transactions on Electron Devices. 2014 Mar 25;**61**(5):1369-1376

[17] Huang JJ, Tseng YM, Hsu CW, Hou TH. Bipolar nonlinear Ni/TiO_2/Ni selector for 1S1R crossbar array applications. IEEE Electron

Device Letters. 2011 Aug 1;**32**(10): 1427-1429

[18] Huang JJ, Tseng YM, Luo WC, Hsu CW, Hou TH. One selector-one resistor (1S1R) crossbar array for high-density flexible memory applications. In 2011 international electron devices meeting 2011 Dec 5 (pp. 31-7). IEEE.

[19] Shin J, Kim I, Biju KP, Jo M, Park J, Lee J, Jung S, Lee W, Kim S, Park S, Hwang H. TiO 2-based metal-insulator-metal selection device for bipolar resistive random access memory cross point application. Journal of Applied Physics. 2011 Feb 1;109(3):033712.

[20] Zhang L, Cosemans S, Wouters DJ, Groeseneken G, Jurczak M, Govoreanu B. Selector design considerations and requirements for 1S1R RRAM crossbar array. In2014 IEEE 6th international memory workshop (IMW) 2014 may 18 (pp. 1-4). IEEE.

[21] Zhang L, Cosemans S, Wouters DJ, Groeseneken G, Jurczak M, Govoreanu B. One-selector one-resistor cross-point array with threshold switching selector. IEEE Transactions on Electron Devices. 2015 Aug 11;**62**(10):3250-3257

[22] Jo SH, Kumar T, Narayanan S, Lu WD, Nazarian H. 3D-stackable crossbar resistive memory based on field assisted superlinear threshold (FAST) selector. In2014 IEEE international electron devices meeting 2014 Dec 15 (pp. 6-7). IEEE.

[23] Srinivasan VS, Chopra S, Karkare P, Bafna P, Lashkare S, Kumbhare P, et al. Punchthrough-diode-based bipolar RRAM selector by Si epitaxy. IEEE Electron Device Letters. 2012 Aug 24; **33**(10):1396-1398

[24] Junsangsri P, Lombardi F. Design of a hybrid memory cell using memristance and ambipolarity. IEEE Transactions on Nanotechnology. 2012 Nov 22;**12**(1):71-80

[25] Kvatinsky S, Wald N, Satat G, Kolodny A, Weiser UC, Friedman EG. MRL—Memristor ratioed logic. In2012 13th international workshop on cellular nanoscale networks and their applications 2012 Aug 29 (pp. 1-6). IEEE.

[26] Kvatinsky S, Belousov D, Liman S, Satat G, Wald N, Friedman EG, et al. MAGIC—Memristor-aided logic. IEEE Transactions on Circuits and Systems II: Express Briefs. 2014 Sep 11;**61**(11): 895-899

[27] Kvatinsky S, Satat G, Wald N, Friedman EG, Kolodny A, Weiser UC. Memristor-based material implication (IMPLY) logic: Design principles and methodologies. IEEE Transactions on Very Large Scale Integration (VLSI) Systems. 2013 Oct 2;**22**(10):2054-2066

[28] Siemon A, Menzel S, Marchewka A, Nishi Y, Waser R, Linn E. Simulation of TaO x-based complementary resistive switches by a physics-based memristive model. In2014 IEEE international symposium on circuits and systems (ISCAS) 2014 Jun 1 (pp. 1420-1423). IEEE.

[29] Amirsoleimani A, Shamsi J, Ahmadi M, Ahmadi A, Alirezaee S, Mohammadi K, et al. Accurate charge transport model for nanoionic memristive devices. Microelectronics Journal. 2017 Jul 1;**65**:49-57

[30] Biolek Z, Biolek D, Biolkova V. SPICE model of memristor with nonlinear dopant drift. Radioengineering. 2009 Jun;**1**:18(2)

[31] Eshraghian K, Cho KR, Kavehei O, Kang SK, Abbott D, Kang SM. Memristor MOS content addressable memory (MCAM): Hybrid architecture for future high performance search engines. IEEE Transactions on Very Large Scale Integration (VLSI) Systems. 2010 May 24;**19**(8):1407-1417

[32] Ho Y, Huang GM, Li P. Dynamical properties and design analysis for

nonvolatile memristor memories. IEEE Transactions on Circuits and Systems I: Regular Papers. 2010 Oct 18;**58**(4): 724-736

[33] Vontobel PO, Robinett W, Kuekes PJ, Stewart DR, Straznicky J, Williams RS. Writing to and reading from a nano-scale crossbar memory based on memristors. Nanotechnology. 2009 Sep 25;**20**(42):425204

[34] Gao Y, Kavehei O, Ranasinghe DC, Al-Sarawi SF, Abbott D. Future large-scale memristive device crossbar arrays: Limits imposed by sneak-path currents on read operations. arXiv preprint arXiv:1507.02077. 2015 Jul 8.

[35] Zidan MA, Fahmy HA, Hussain MM, Salama KN. Memristor-based memory: The sneak paths problem and solutions. Microelectronics Journal. 2013 Feb 1;**44**(2):176-183

[36] Fei W, Yu H, Zhang W, Yeo KS. Design exploration of hybrid cmos and memristor circuit by new modified nodal analysis. IEEE Transactions on Very Large Scale Integration (VLSI) Systems. 2011 May 5;**20**(6):1012-1025

[37] Maeda N, Komatsu S, Morimoto M, Tanaka K, Tsukamoto Y, Nii K, et al. A 0.41 µa standby leakage 32 kb embedded sram with low-voltage resume-standby utilizing all digital current comparator in 28 nm hkmg cmos. IEEE Journal of Solid-State Circuits. 2013 Jan 15;**48**(4): 917-923

[38] Lehtonen E, Poikonen JH, Laiho M, Kanerva P. Large-scale memristive associative memories. IEEE Transactions on Very Large Scale Integration (VLSI) Systems. 2013 Apr 30;**22**(3):562-574

Chapter 7

Functional Capabilities of Coupled Memristor-Based Reactance-Less Oscillators

Vladimir V. Rakitin and Sergey G. Rusakov

Abstract

New functionalities of reactance-less memristor based oscillators are discussed which arise when two elementary oscillators are connected. It is shown that the system of coupled memristor based oscillators can be used for converting analog and analog-digital signals into binary pulse sequences. The approach to control the thresholds in memristor based oscillators is discussed. Standard control approach in memristor based oscillators is the exploitation of input signal to drive the rate of change in the state of the memristor. In contrast, the main idea of the considered controlling approach is to send the input signal not directly to the memristor device but to the comparator circuit and as result to control oscillator circuit behavior by change of interval of memristor resistor variation. The capabilities of coupled memristor based oscillators with control thresholds are sufficient for constructing the simple circuit elements of oscillatory computing architectures.

Keywords: reactance-less memristor based oscillators, coupling oscillators, memristor devices, threshold comparator, switching thresholds, binary oscillator networks

1. Introduction

The simplicity of the design of memristor based circuits and the possibility of manufacturing memristors [1–3] using integrated technology make them promising for use in a variety of information storage and processing systems. The construction of neuromorphic systems [4–8] is one of the most important memristor applications where the memristors provide the function of nonvolatile analog memory.

Due to memristor capabilities the wide implementation of memristors is predicted in different circuit application spheres including analog circuits. The properties of memristors [3, 9] open up new possibilities of constructing the memristor based oscillators (MBO) of different types [10–14]. The complex behavior of MBOs is analyzed in some papers (see for instance [15–18]). The inertial property of memristors provides the elimination from oscillator circuits the reactive elements (inductors and capacitors) which are poorly compatible with the requirements of the integrated implementation of neuromorphic systems. By the present time the various types of reactance-less MBO have been proposed [19–28]. This class of oscillators is considered below in the paper.

The neuromorphic systems including artificial neurons (AN) and networks become promising area where the analog memory plays the important role [29–39].

The memory elements are located between neurons and provide restructuring the coupling weight coefficients. Memristors are well suited to the requirements for artificial synapses [9, 40, 41]. The memristor resistance determines the value of the weight coefficients. The change in resistance under the action of current determines the possibility of restructuring the connections.

However, it should be noted that the properties of memristors allow them to be used not only as synaptic elements but also in the artificial neurons themselves. It can be mentioned that the reactance-less MBO consisting of memristor device and an active element, for instance comparator, can be also considered as simple AN model. Such an oscillator element can be inhibited or excited similarly to AN behavior. Its state can be specified by the phase of periodic oscillation.

Advanced AN models [8] that more accurately describe the behavior of biological neurons have high complexity to represent essentially more complex and various dynamical processes. The response of oscillatory AN to the input excitation involves not only changing the state but also changing the character of generation of output pulse train. In this case the number of the pulses and position of the pulses in pulse train depend on input amplitude and transient prehistory.

The complex mathematical model is required to represent such a behavior. This is usually achieved by increasing the order of the model. The complexity of circuits of corresponding oscillatory AN is also must be increased [42, 43] and strict requirements for the precision of circuit parameters must be met.

We present the alternative approach in this paper. We demonstrate that coupled memristor-based reactance-less oscillators have the set of modes with dynamical processes that is enough to provide the desired complex behavior. To support these capabilities at circuit level the approach to MBO construction is presented that based on controlling the comparator threshold. Some advantages of this approach are demonstrated.

Among the advantages of controlling threshold approach in MBO it is essential to point out the opportunity to construct piecewise constant (PWC) oscillators. Recently AN models based on piecewise constant (PWC) oscillators have appeared [44–46]. Such AN models are convenient in practice. PWC oscillators are the oscillators with mathematical models which are systems of ordinary differential equations (ODE) with piecewise constant coefficients. The signals generated by AN in this case are piecewise linear functions of time. PWC oscillators are developed on the base of standard electronic components including amplifiers, logic gates, resistors, capacitors. The transient processes occur in these circuits under constant excitation, for example the charge or discharge of the capacitor at constant current. The analysis of AN behavior of such type and networks based on them is given in papers [47, 48]. The nonlinearity of the memristor characteristics due to the change in its resistance when current flows through device limits the development of PWC memristor based oscillators [49, 50]. Application of the considered approach to control threshold in MBO avoids this restriction because it provides use only changing the sign of the current through the memristor while generation process.

Application in binary oscillator networks is other important capability of the considered coupled reactance-less MBOs. Oscillatory neural networks are promising candidates for solving a number of complex computational problems [51–55]. The most suitable circuit elements for such networks are binary generators with binary output signals [56–58]. In binary oscillator networks (BON) binary signals are exchanged and information is represented by binary streams. The considered coupled reactance-less MBOs can be applied as elementary binary oscillators.

The rest of the paper is organized as follows. Section 2 presents the principle of controlling thresholds in MBO circuits. The circuit version of coupled MBOs with positive couplings and its functionalities are discussed in Section 3. In Section 4 the functional capabilities of coupled MBOs with inverting connections are given. The main properties of coupled MBO for use in binary generator networks are considered in Section 5. The technique of using phase planes to analyze the behavior of MBOs is widely used in sections.

2. Foundation: the principle of controlling threshold parameters in memristor based oscillator circuits

2.1 Operating principles of reactance-less memristor based oscillators

Oscillators without inductors and capacitors are the result of the memristor features applying. The self-excitation conditions are provided by the inertia of the resistance change of memristors when current flows through memristors devices. The absence of reactive elements allows to minimize the size of memristor based oscillators (MBO). The requirements to oscillator-based computing are met, in particular, by various variants of MBOs that differ in the number of memristor devices and the techniques of their coupling.

The schematic of typical reactance-less MBO is shown in **Figure 1a**. The circuit consists of memristor device M and a two-threshold comparator (TTC) with a current generator. The comparator converts the voltage v on the memristor to a binary output signal v_{out} (**Figure 1b**). The current generator converts the output binary signal ("0" and "1") of v_{out} to opposite corresponding currents $i(v)$ ($-I$ and $+I$). The current input i_{in} is conventional input for reactance-less MBO. The memristor is connected to the input of the comparator by anode.

The memristor resistance R is decreased at a positive voltage v at anode when a positive current i flows in. The transfer function of the comparator is shown in **Figure 1b**. The comparator output voltage is "0" at $-V_M < v < V_m$ and it is equal to

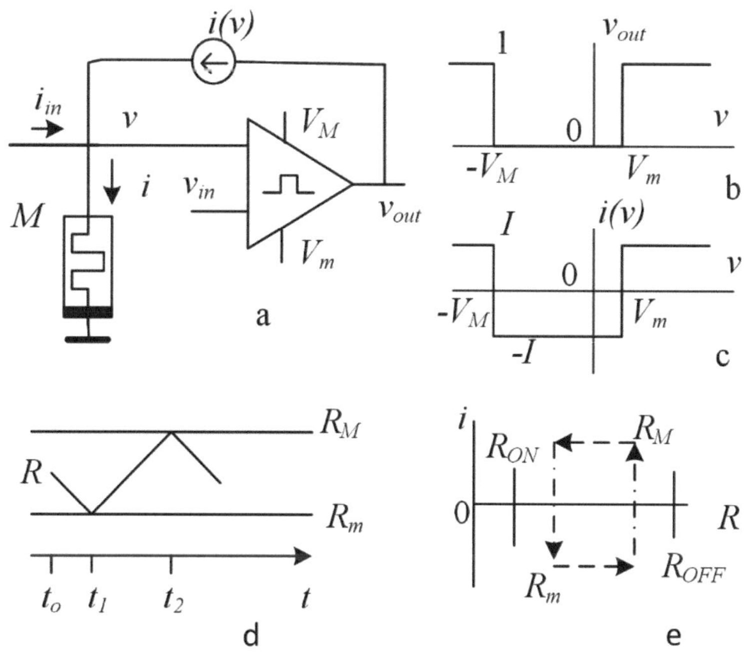

Figure 1.
Typical illustrative graphs of behavior of reactance-less memristor based oscillator: (a) schematic of memristor based oscillator), (b) transfer function of comparator, (c) input function of comparator with current source, (d) waveforms of varying memristor resistor, (e) hysteresis loop for memristor resistor at phase plan.

"1" otherwise. Here $V_M > V_m$. The current generator in the negative feedback circuit of the comparator converts the binary output signal ("0", "1") into a negative current and a positive current through the memristor $(-I, +I)$, respectively (**Figure 1c**). The input current i_{in} is summed with the current $i(v)$.

The memristor resistance can be considered as characteristic of oscillator state. Typical graph of varying memristor resistance in self-excitation mode of oscillator is given in **Figure 1d**. The phase plan (**Figure 1e**) illustrates the cycle of change of the memristor resistance R while oscillations as hysteresis loop.

Let us consider the cycle of periodic self-excitation mode of memristor oscillator (**Figure 1a**). Let's assume that for the initial moment of time t_0 the voltage value v is $v > V_m$ (**Figure 1c**). In this case the current is positive $i = I$ and $v_{out} = "1"$. Therefore, the memristor resistance R and the memristor voltage are reduced. At time t_1 the voltage reaches the threshold value $v = V_m$, the output voltage v_{out} goes from state "1" to state "0". The value of memristor resistance is $R_m = V_m/I$ at this time point. Here R_m is lower threshold value of the memristor resistance. In this case current i and voltage v become negative: $i = -I$ and $= -V_m$. The memristor resistance begin to increase, this leads to decreasing the negative voltage on the memristor. At time t_2 it reaches the value $v = -V_M$, the output of the comparator goes from "0" to "1", the current and voltage on the memristor become positive again: $i = I, v = V_M$. At this time point, the resistance of the memristor achieves the value $R_M = V_M/I$ where R_M is upper threshold resistance value. To provide periodicity of this process the following conditions must be satisfied

$$R_{ON} < R_m = \frac{V_m}{I} < \frac{V_M}{I} = R_M < R_{OFF}. \tag{1}$$

Here R_{ON} - is the minimal memristor resistance, R_{OFF} – is the maximal memristor resistance. In this case, the memristor resistance will periodically change in the range from the lower threshold value R_m to the upper threshold resistance R_M (**Figure 1d**). The change in resistance is triangular if the rate of change in the memristor resistance does not depend on its value. The rate of change is proportional to the current according to the drift-diffusion model approximation [3].

The input current impacts on the speed of memristor resistance change. The speed is increased at the same signs of the input current and the generator current and it is decreased in opposite case.

2.2 Introducing the control of threshold parameters in memristor based oscillator circuits

Standard control approach in memristor based oscillators (MBO) is the exploitation of input signal to control the rate of change in the state of the memristor.

In contrast from this, the main idea of considered controlling approach is to send the input signal not directly to the memristor device but to the comparator circuit and as result to control oscillator circuit behavior by change of interval of memristor resistor variation.

The possible schematic of memristor based oscillator with controlled threshold parameters [50] is given in **Figure 2**. This oscillator element provides the desired functionalities.

The purpose is to change the comparator thresholds using the input voltage V_{IN} and to control the boundaries of range of memristor resistance variation by input voltage. In this case input voltage $V_{IN}(t)$, limited by the region $V_{OUT}(t) \geq V_{IN}(t) \geq 0$, shifts the range of $R(t)$ change:

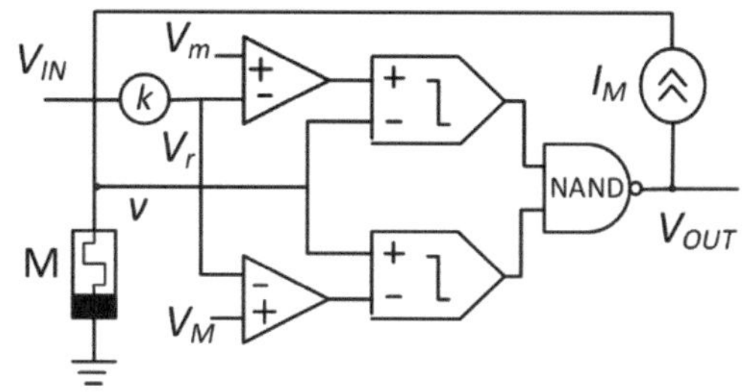

Figure 2.
Schematic of memristor based oscillator with controlled threshold parameters. The oscillator circuit contains memristor M, *two-threshold comparator (TTC), summing elements, attenuator* k $(V_r = kV_{IN})$, *current source* I_M., *logical element* NAND.

$$R_m - r(t) \le R(t) \le R_M - r(t). \tag{2}$$

Here $r(t) = V_r(t)/I$ is conditional resistance. In order to avoid exceeding the limits of the range of changes in the memristor resistance, the following inequalities are supported:

$$r(t) < R_m - R_{ON} \text{ and } r(t) < R_{OFF} - R_M \tag{3}$$

The original comparator thresholds V_m and V_M are converted into active thresholds in this case.

It can be mentioned that the state of the MBO can be characterized by phase. The phase is determined by the values of two variables: $R(t)$ and $sign(dR/dt)$.

The fundamental difference between the proposed control approach and the conventional approach is following: the change of the memristor state does not depend on the time of the drive signal arrival under standard control and the state change depends on the time of arrival of the drive signal for proposed approach.

The different character of impact of driving pulses on MBO behavior is shown in **Figure 3**. The input current i_{in} impacts on the rate of change in the memristor resistance (**Figure 3a**). In this case the speed increases at the same signs of the input

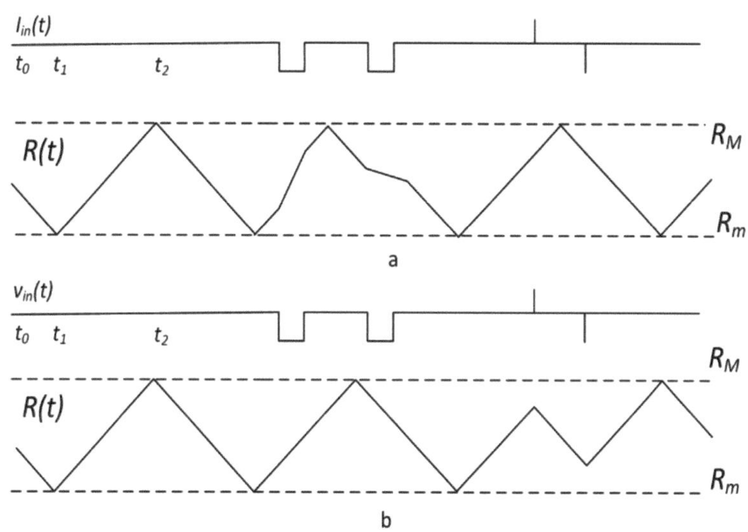

Figure 3.
The different character of impact of driving pulses on varying memristor resistance R(t)*: (a) excitation by input current* i_{in}*, (b). excitation by input voltage* v_{in} *to control thresholds.*

current and oscillator current and it decreases otherwise. The input voltage V_{IN} is applied to the comparator to change its thresholds and to determine the range of resistance changes (**Figure 3b**).

The input current signal i_{in} is integrated. Its effect on the waveforms depends on the duration of the signal and on the phase of the process, in other words on the sign of the resistance change. The long current pulses slow down or accelerate the transient process but short current pulses do not impact on the resistance value.

The input signal V_{IN} applied to the input of the comparator directly before reaching the threshold can affect the switching process even with a small value of the coefficient k. At other times the comparator sensitivity to the input signals is reduced. This is illustrated in **Figure 3b**. The long-time pulses applied to the comparator input do not affect the trajectory $R(t)$. But even a short positive pulse before reaching the upper threshold resistance R_M leads to a decrease in the switching threshold and to earlier start of reducing memristor resistance. Similarly, a short negative pulse before reaching the lower threshold resistance R_m leads to an increase in the lower switching threshold. This leads to beginning of growth of the memristor resistance.

2.3 Applicability of memristor based oscillator circuits with control of thresholds in oscillator networks

The pointed out features of two considered approaches to control MBOs predefine their exploitation in oscillator networks. The current inputs are more suitable for controlling the state of network elements by external signals. The voltage inputs with control of thresholds should be used to organize interaction of network elements with each other including synchronization mode of oscillators.

Then we will limit ourselves to the consideration of MBOs with voltage inputs and corresponding control of thresholds. Such an oscillator element can be considered as binary element with the binary input v_{in} (**Figure 4**).

The current is positive and the memristor resistance decreases at the output signal $v_{out} = "1"$, until the resistance reaches the lower threshold $R_m - r = R_m - kv_{in}/I$. At output signal "0" ($v_{out} = "0"$) the current through the memristor is negative, its resistance increases until it achieves the upper threshold $R_M - r = R_M - kv_{in}/I$.

Thus, input state $v_{in} = "1"$ slows down the exit from the state $v_{out} = "1"$, and accelerates the exit from the state $v_{out} = "0"$.

It can be mentioned that considered MBOs with voltage control of thresholds are well suitable for synchronization mode of coupled oscillators due to high sensitivity to external input and fast transient to synchronization steady state.

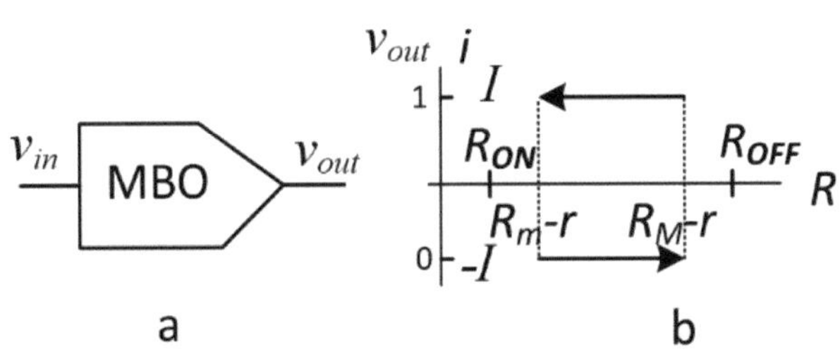

Figure 4.
Binary MBO element (a) and hysteresis loop for MBO resistor (b) at phase plan taking into account threshold shift.

2.4 Model equations

The linear drift model [3] can be applied to describe the behavior of the memristor device. This model involves "instant" voltage–current characteristic for the instantaneous value of resistance

$$v = R \cdot i \tag{4}$$

and control characteristic given by differential equation:

$$\frac{dR}{dt} = -\mu \frac{(R_{OFF} - R_{ON})R_{ON}}{D^2} i = -\gamma i, \tag{5}$$

The model has the following parameters: the high memristor resistor value R_{OFF}, the low memristor resistor value R_{ON}, ion mobility μ, the semiconductor film thickness D, γ-is inertial parameter. These parameters have the following typical values:

R_{OFF} = 10 kOhm, R_{ON} = 1 kOhm, $\mu = 10^{-14}$ m^2 s^{-1} V^{-1}, D = 10 nm [3], $\gamma = 10^9$ V A^{-2} s^{-1}.

The switching time of the memristor device under constant current I can be estimated as

$$T_R = \frac{R_{OFF} - R_{ON}}{\gamma I} \approx \frac{R_{OFF}}{\gamma I}, \tag{6}$$

This time is 100 ms for current value 100 μA.

To describe the behavior of oscillator with comparator it is convenient to exploit the dimensionless parameters and variables. The dimensionless time is also applied. Such dimensionless variables can be obtained by normalizing. The normalization of resistances is performed using division by R_{OFF}, respectively for voltages division by $(I \cdot R_{OFF})$ is applied and for time - division by T_R. As a result we have $R_{OFF} = 1$, $R_{ON} - 0.1$ and γ =1.

Taking into account the threshold shift the comparator model with current generator $i = i(v)$ (**Figure 1a**) is described by the equations

$$i(v) = \begin{cases} 1, if\ v > V_m - kv_{in} \\ -1, if\ -(V_M - kv_{in}) < v < V_m - kv_{in} \\ 1, if\ -(V_M - kv_{in}) > v \end{cases} \tag{7}$$

Taking into account the binary variables the equations Eqs. (4), (5), and (7) can be transformed to single piecewise constant equation with respect to the variable R

$$\frac{dR(t)}{dt} = \begin{cases} -1, if\ R(t) > R_M - r(t), \\ 1, if\ R(t) < R_m - r(t), \\ \dfrac{dR(t - \Delta\tau, \Delta\tau \to 0)}{dt}, if\ R_M - r(t) > R(t) > R_m - r(t), \end{cases} \tag{8}$$

where $r(t) = kv_{in}(t)/I$ –as mentioned above, variable that reflects the change in the threshold under the influence of external signal. To save the oscillation conditions the following restrictions for $r(t)$ must be satisfied:

$$R_{OFF} > R_M + r(t), R_M - r(t) > R_m + r(t), R_m - r(t) > R_{ON} \tag{9}$$

The solution of equation Eq. (8) has character of triangular oscillations in the range $R_m - r < R(t) < R_M - r$.

2.5 Features of reactance-less memristor based oscillators in low frequency applications

The model equation Eq. (5) describes an important feature of memristors - the property of inertia. Due to this property it is possible to construct the reactance-less oscillators or in other words oscillators without inductors and capacitors. In this case the charge and discharge of reactive components in conventional oscillators is replaced by changing the memristor resistance (Eq. (5)). The duration of these processes is determined by the inertial parameter γ. The typical times of switching of the memristor devices are determined by Eq. (6). Thus, typical current value 100 μA corresponds to oscillator frequency 10 Hz.

The prospects of application of such oscillators are associated primarily with the development of low-power low-frequency oscillator circuits for neuromorphic systems and biomedical equipment.

The low-frequency operation range is the main application area of memristor oscillators [21]. Low frequency oscillators are important for many applications but their design is connected with significant difficulties due to the large values of capacitors required for low oscillation frequencies [59]. Since the frequency of operation of conventional RC relaxation oscillators is inversely proportional to the time constant, $\tau = R \times C$, low-frequency operation requires high capacitance [21]. In this case the typical capacitance value may exceed 1 μF, capacitor occupies an area of more than mm^2. Such an area size contradicts the implementation in integrated circuits. This leads often to off-chip placement of the capacitor [21]. The special-purpose techniques are developed to overcome this problem and to avoid the use of impractically large component values [59, 60]. Thus, relatively novel technique was used to implement the oscillator on-chip, but the capacitor consumed 77.8% of the total chip area [60].

It can be mentioned that the problem is solved automatically with applying reactance-less MBOs due to very small area of memristor devices. For the considered MBO circuits of type (**Figure 2**) the size of area is determined by the area occupied by CMOS comparator.

2.6 Alternative circuitry

In this type of MBO the comparator plays the role of control circuit in switching the direction of the memristor current. It can be noted that this function can be performed by other active circuit elements.

In particular, circuit with a series connected two devices can be considered: memristor and device with negative differential resistance (NDR). This circuit can generate relaxation oscillations when the generation conditions are satisfied.

There is no need for an active load in such circuits. This is advantage of oscillator circuits based on memristor with NDR. In particular, such two-terminal devices can specified by S-shaped I-V characteristics. In this case the memristor itself can have two state given by high and low resistance values [35]. The relaxation oscillations become possible when memristor is connected to a passive two-terminal circuit. Such oscillators can be connected to each other by resistive or resistive-capacitive couplings. This type of oscillators corresponds to circuits with the current input.

Various two-terminal devices can be used as the load in oscillators based on memristors with NDR. Among them, devices with a structure similar to memristors that exploit thin layers of insulators are promising. Creation of such devices based on silicon oxides [36, 61] seems to be the most promising now. New emerging memristive technologies such as SiOx-based memristors are discussed in [61]. The compatibility with standard CMOS technology provides a good perspective for the implementation of hybrid CMOS-memristive designs in various applications.

Recent results [61] demonstrate advantages of the architecture of memory cell comprising memristor and selector. It is expected that under certain conditions such an emerging device architecture can act as an oscillator.

In the following text the consideration is limited by oscillator circuits based on memristor devices [3], although the results presented below for coupled oscillator elements can be extended to above mentioned circuit architecture.

3. Behavior of coupled memristor based oscillators with positive couplings

3.1 Operating principles

The analysis of behavior of two coupled identical MBO with positive connection is presented below.

This circuit is shown in **Figure 5**. It contains MBO1, MBO2, an adder at the input and a phase detector at the output [50]. To provide an external control the excitation signal V_C is transmitted using an adder at the input. The phase detector at the output is used to identify the synchronization mode of coupled oscillators. If there is no synchronization between the oscillator stages MBO1 and MBO2 then output signal $V_s = 1$ and $V_s = 0$ if there is synchronization.

The coupling strengths between the MBOs specified by coefficient k impact on the behavior of this system significantly.

The rates of change of memristor resistances R_1 and R_2 are equal in modulus for identical MBOs. But these rates may differ in signs. By such a way the variables R_1 and R_2 and the signs of derivatives dR_1/dt and dR_2/dt can be considered as system states and may specify the behavior of system of two coupled oscillators.

The phase plane with axes R_1 and R_2 (**Figure 6**) can be exploited for analysis of different behavior versions of such a system. The analysis is based on model Eq. (8). In this case, the trajectories of moving the image points are straight lines. They pass at angles of $\pm \pi/4$ on phase plane. Four trajectories can pass through each point of phase plane. The sign of dR/dt defines one from them.

The boundaries of the area of trajectories movement are specified by the threshold resistances. When the trajectory reaches the boundary the sign of the derivative dR/dt changes and trajectory is mirrored from the boundary. The boundaries can shift themselves at this time point.

If the external excitations are absent then the threshold of each MBO depends on positive pulse from the neighboring MBO. In particular the lower limit of the resistance of each MBO is reduced to $R_m - r$. As a result the area of the allowable system states on the phase plane in self-oscillating mode is determined by the square with vertices (R_M, R_M) and $(R_m - r, R_m - r)$ (**Figure 6**). The area of

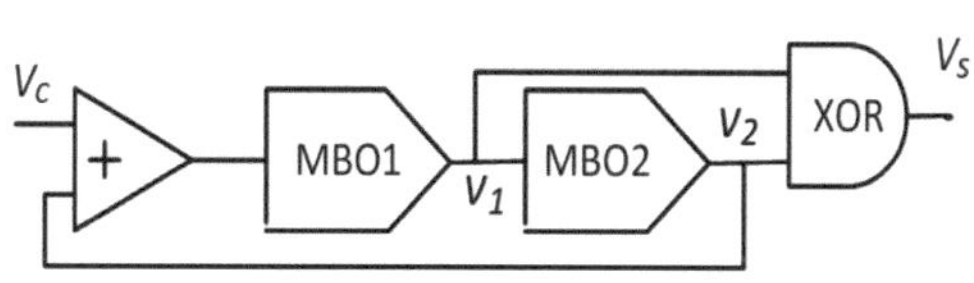

Figure 5.
Schematic of coupled memristor based oscillators (MBOs).

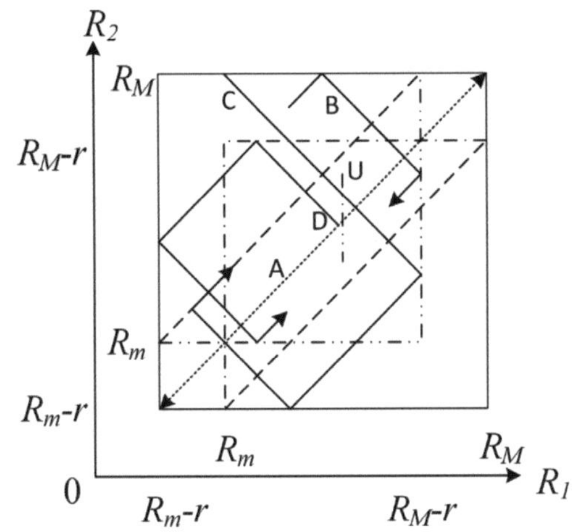

Figure 6.
The boundaries and trajectories at phase plane of changing the variables R_1 and R_2 for coupled MBOs: solid lines – boundaries for case $dR_1/dt = dR_2/dt$, dash-dotted lines - boundaries for case $dR_1/dt \neq dR_2/dt$, solid lines with arrows - the trajectories of R_1 and R_2 with different initial conditions. The areas of stable trajectories are limited by dashed lines.

stationary trajectories is located insight this square. This area is limited by the dashed lines in **Figure 6**.

For the existence of a stationary trajectory, the following necessary and sufficient conditions must be met: the image points must be located in the area indicated above, and the signs of the derivatives must be identical.

If the variables are located at the main diagonal in this area and the specified conditions are met, then the variables reach the threshold simultaneously (dotted line A in **Figure 6**). Their moving directions also change simultaneously. They continue to move along the main diagonal. When the threshold line is reached by one variable on the other lines parallel to the main diagonal in this area, the sign of its derivative changes. This is followed by the threshold change for another variable with a corresponding change in the sign of its derivative. The trajectory is saved, but the movement along it occurs in the opposite direction. Note that the phases of the oscillations of the resistors are the same ($V_s = 0$) for stable trajectories.

If the starting points of trajectories are located outside area of stationary trajectories (**Figure 6**) then such trajectories are reflected after reaching the boundaries. If in this case the signs of the derivatives are the same then the segments of the trajectories tend to the stability region. The reflection character is defined by the boundaries with different signs of derivatives $dR_1/dt \neq dR_2/dt$ (dashed lines in **Figure 6**). Any trajectory ends in the region of stable trajectories in result. Such behavior is illustrated in **Figure 6** by examples of the trajectories B and C. It can be seen that the trajectory B falls into the stability region after two reflections and the trajectory C - after four reflections.

The considered circuit with two coupled identical oscillator elements (**Figure 5**) has a set of stable and unstable steady state trajectories. The difference between the maximal values of the variables $R_s = R_{1max} - R_{2max}$. can be exploited as characteristic of stable steady state trajectories. It can be mentioned that zero value $R_s(R_s = 0)$ corresponds to the main diagonal on phase plan (**Figure 6**). This characteristic reaches the value $R_s = \pm r$ at the boundaries of the stable region. The each stationary trajectory (each value of R_s) corresponds to a certain period of triangular oscillations which equals to

$$T_S = 2\frac{R_M - R_m + r - R_S}{\gamma I}. \tag{10}$$

Let duration of the additional external control signal V_C be shorter than period T_S. This signal V_C can change the boundary and the trajectory of movement on the phase plane respectively. **Figure 6** shows the boundary U created by an external signal. The trajectory D in **Figure 6** illustrates the transition to new stable trajectory under the influence of an external signal. The starting point of trajectory D is located at the main diagonal. The trajectory D moves away from the main diagonal under the external excitation. After three reflections (**Figure 6**) the transition of image point to new stable trajectory is carried out.

3.2 Features

It can be mentioned that for considered coupled MBOs the movement along the trajectory in the direction opposite to the original one can be provided by changing the signs of the derivatives. This property can be called as reversibility of trajectories. The property is valid for stable trajectories as well as for any unstable trajectories before its transition to stable ones. Such a feature may be foundation for the management of coupled MBOs.

In order to get from the original fixed trajectory (for example A) onto given trajectory (for example D), it is enough to choose the intersection point of the predetermined path with the threshold line ($R_2 = R_m$) and then to construct the trajectory of leaving it until the inevitable intersection with the original trajectory using change in derivative sign. The control signal with short duration and sufficient amplitude moves the image point to the specified trajectory.

The process of transition to stationary trajectory can be represented using the mapping function of the value R_S over the period: $R_S(n+1) = P(R_S(n))$ (**Figure** 7). The value of R_S for the *n*-th period is given in **Figure** 7 at the abscissa axis and similar value as a result of Poincare mapping for *n+1* period is shown at the ordinate axis. The area of stable states belonging to the diagonal D (**Figure** 7) satisfies the condition: $-r < R_s < r$, that corresponds to the area of stationary trajectories.

Until $|R_S| > 3r$ the return of R_S to the region of stable states is performed with stepsize equal to $4r$. If R_S located in the interval $r < R_S < 3r$ then the return occurs in one step equal to $2(|R_S| - 1)$.

As follows from this analysis, the speed of the transition process from the excited state to the stationary state depends on the coupling strength or in other words on the coefficient r. The width of the stability area also depends on coupling strength. The return to the stationary trajectory after external excitation can be relatively long at low values of factor r. It can be expected that the return time is

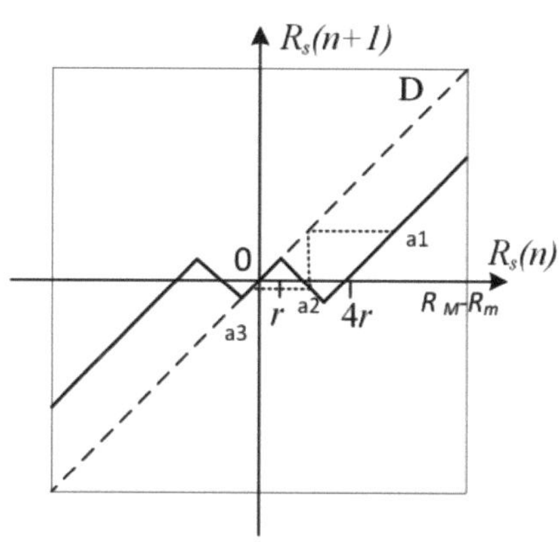

Figure 7.
The function of mapping the difference in the states of the coupled MBOs for period.

proportional to the amplitude of the external signal in a certain range of amplitude varying.

The situation changes significantly when the excitation has a long duration, comparable to or exceeding the duration of the period T_S. In this case new values of the threshold resistances for R_1 are set during the action of the input signal. Then the value R_1 will change within this interval of variation. The duration of the transition to the perturbed state will also depend on the coupling strength. The output signal will appear on the phase detector in this case. The return to the stationary trajectory will repeat again after completion of the input signal and signals at the output of the phase detector will appear again.

The behavior of self-oscillating coupled MBO is described by piecewise-constant differential equations. As a result, the complete analytical solution can be obtained. In practice, it reduces to solving the problem of elastic reflection of a point inside a rectangle with edges positioned depending on the sign of the point's speed.

3.3 Simulation examples

Below the results of simulation of the coupled MBOs are given. The simulation examples demonstrate the opportunity to control the state of the coupling MBOs and illustrate also waveforms of generation of the pulse trains at the input excitation.

The time is defined as dimensionless variable. Also, the dimensionless values of the circuit parameters and variables were used during simulation. Among them: $R_M = 0.8$, $R_m = 0.4$, $r = 0.1$.

3.3.1 Example 1: managing the state of coupling MBOs

The considered circuit example has a set of stable and unstable steady- state trajectories and provides complex transformation of input signal. The simulation example illustrates the presence of three stable steady-state periodic solutions (**Figure 8**) correspond to $R_s = R_{1max} - R_{2max} > 0$, $R_s = R_{1max} - R_{2max} < 0$, $R_s = R_{1max} - R_{2max} = 0$.

Input signals lead to switching of stable trajectories and provides various modes in application.

Let the starting points for the variable resistances be the same al for MBO1 and MBO2 ($R_1(0) = R_2(0) = 0.5$). When the first control pulse with amplitude $V_C = 0.4$ and duration $T = 0.5$ is applied to MBO1 the process is generated in which the amplitude of the resistance oscillations of MBO1 is greater than the similar amplitude for MBO2. The state corresponds to inequality $R_S > 0$ (**Figure 8**). The second

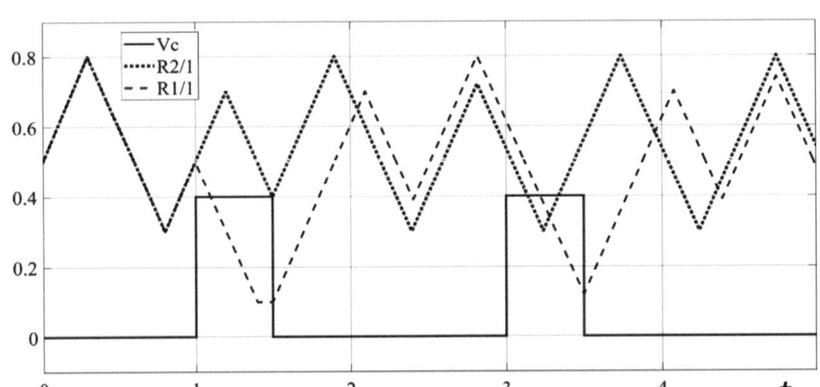

Figure 8.
The computed waveforms in the coupled MBOs. V_C - the solid line, $R_1(t)$ - dashed line, $R_2(t)$ - dotted line.

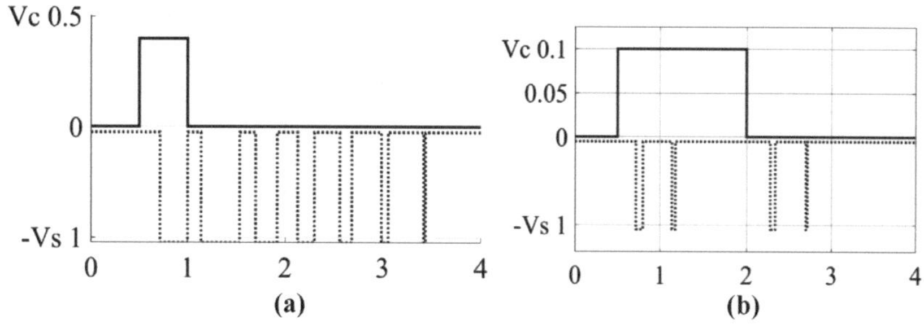

Figure 9.
Generation of train of output pulses (dotted line) under the action of the input signal of short duration (a) and long duration (b).

control pulse leads to change of trajectory and generates the steady state corresponding to inequality $R_S < 0$ (**Figure 8**).

By such a way this simulation example confirms the predicted change of memristor states in the considered circuit under control pulse excitation.

3.3.2 *Example 2: generation of pulse train by coupling MBO under the action of the input signal*

Let the initial setting conditions for MBO1 and MBO2 be the same that corresponds to the zero voltage V_S at the output of the detector (**Figure 9a**). After the input signal with amplitude $V_C = 0.5$ and duration T = 0.5 (**Figure 9a**) their synchronization is violated for time proportional to the amplitude of the input action. The output signal of the comparator with amplitude $V_S = 1$ appears after the beginning of the transition to the perturbed trajectory and pulses remain for a long time.

When long-term input signal T=1.5 of relatively small amplitude $V_C = 0.1$ is applied the transition to the perturbed trajectory and exit from it is performed in shorter time (**Figure 9b**). The output signal $V_S = 1$ occurs both after the rising slope and after the fall slope of the input pulse.

3.4 Output

The coupled memristor based oscillators with positive couplings have a set of stationary states in self-excitation mode.

An external signal can initiate a transition from one stationary state to another. Also such a signal can remove the system from the region of stationary states to the excited mode. This excitation is saved after completion of the input signal. The transition to new steady state takes some time after completion of the external excitation. The pulse train is generated at the comparator output during this time interval.

The coupling memristor based oscillators can be considered as the analog-to-digital converters that provide conversion of input amplitude variation.

4. Behavior of coupled memristor based oscillators with inverting connections

4.1 Operating principles

The connection types of the coupled memristor based oscillators (MBO) and the values of the coupling strengths between them impact significantly on the character of their behavior.

The schematic of coupled MBOs with inverting connections and binary output signals is shown in **Figure 10**. The circuit of this oscillator element contains [49] two identical oscillators MBO1 and MBO2. If the direct signal V_1 applied to input MBO2 then MBO1 receives an inverted signal $(-V_2)$ from MBO2 output. The circuit contains also output phase detector $F(V_1, V_2)$ and input adder. The phase detector performs logical function over the binary outputs MBO1 and MBO2. The input adder provides the receipt of both the control analog signal V_C and the inverted signal $(-V_2)$ at MBO1 input.

The state of the considered oscillator system can be specified by the variables R_1 and R_2 and time derivatives dR_1/dt and dR_2/dt. The modules of the rates of change of memristor resistances R_1 and R_2 are the same for identical MBOs but signs of these rates may differ.

The detailed analysis of behavior of this system using phase plane for variables R_1 and R_2 is given in [49]. The feature of phase portrait for coupled MBOs with inverting connections is related with the derivatives of variables R_1 and R_2 that take the values ± 1. Due to this feature the trajectories of the image point are inclined straight lines with angle of $\pm\ \pi/4$ relative to the coordinate axes.

The ratio of coupling coefficients with opposite signs $r_1 = -k_1 V_2(t)/I$ and $r_2 = k_2 V_1(t)/I$ significantly affects character of behavior. The cases of equal values $(|r_1| = r_2)$ and different values $(|r_1| \neq r_2)$ are discussed in [49].

Introducing the additional notations r and r_M two possible versions can be considered for different values of coupling coefficients: $|r_1| = r_m < r = r_2$, $|r_1| = r_M > r = r_2$. First case with the dominance of direct positive coupling corresponds to antiphase oscillations and second case with dominance of inverting negative coupling corresponds to in-phase oscillations.

The period of antiphase oscillations equals to $T = 2\,(R_M - R_m - r_m)/\gamma I$ and the period of in-phase oscillations is $T = 2\,(R_M - R_m - r)/\gamma I$.

The external control signal impacts on the phase trajectory of the system. Consider then the case with $|r_1| = r_m$.

The range of varying MBO1 threshold voltage is shifted due to applying the control signal V_C. The additional shift in the threshold resistances $r_C = V_C/I$ is generated by control signal V_C . Due to action of the V_C signal the following active restrictions determine the interval of varying memristor resistances of MBO1 and MBO2 circuits.

$$R_m + r - r_C \leq R_1(t) \leq R_M + r - r_C \; at \; dR_2/dt < 0 \tag{11}$$

$$R_m - r_C \leq R_1(t) \leq R_M - r_C \; at \; dR_2/dt > 0 \tag{12}$$

$$R_m - r \leq R_2(t) \leq R_M - r \; at \; dR_1/dt < 0 \tag{13}$$

$$R_m \leq R_2(t) \leq R_M \; at \; dR_1/dt > 0 \tag{14}$$

Figure 11 illustrates such shift at the phase portrait of the system with control signal. As follows from formulas Eqs. (11) and (12), the threshold resistances for R_1 are decreased (**Figure 11**) but the threshold resistances for R_2 remained unchanged.

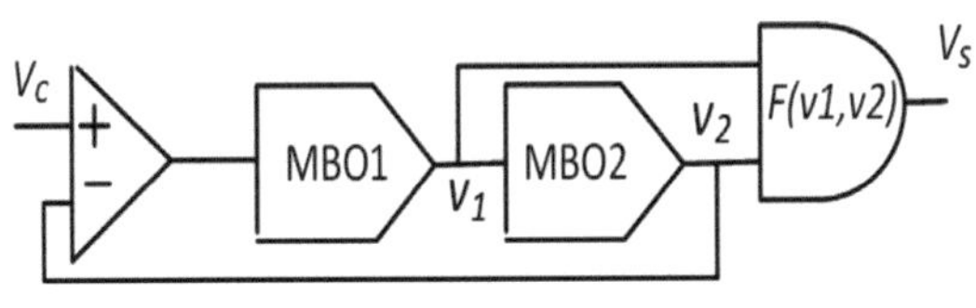

Figure 10.
The system of coupled MBOs with inverting connections.

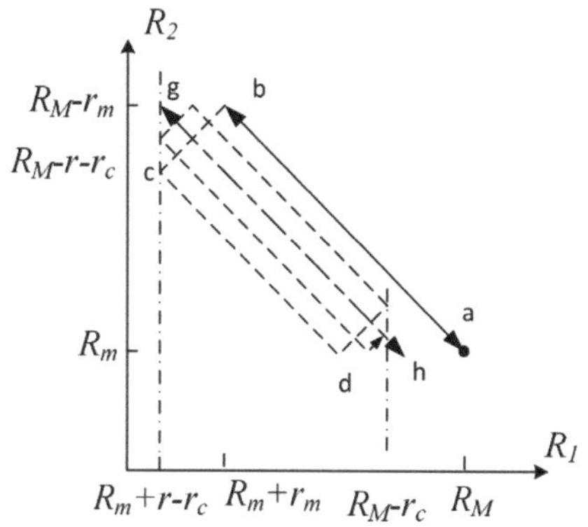

Figure 11.
Phase plane of system of coupled MBOs with inverting connections under external excitation.

For this reason, the parallel shift of trajectories at the phase portrait (**Figure 11**) corresponds to impact of external control signal.

Let the initial stable trajectory of the system before an external excitation correspond to line segment *(ab)* and after an external excitation the displaced trajectory corresponds to line segment (*gh*). When a constant control signal is applied for sufficiently long time, the transition to the trajectory (*gh*) is inevitable. It is caused by change in the sign of dR_2/dt and reducing the threshold resistance to $R_m + r - r_C$ when the point b reaches the border at $R_1(t) = R_m + r_m$. **Figure 11** illustrates the movement of image point from b to c, then to d, until it falls on the trajectory *(gh)*.

The difference $\Delta = r_2 - |r_1|$ determines the width of the stability area and impacts on the speed of the transition process to new trajectory. It should be expected that due to the piecewise linear character of transients the return time will be proportional to the input amplitude in certain range of amplitude variation of control signal.

4.2 Simulation example

The results of the behavior simulation of the coupled MBOs with inverting connections are given below for case of short input signal.

The computed waveforms for the oscillator system with phase detector NOR are shown in **Figure 12**. In this case stable antiphase oscillations are observed in the system under the absence of an external signal.

The following values of coupling factors were selected: $r_2 = 0.1, r_m = 0.09$. These values of coupling factors mean that positive coupling in connected MBOs is stronger than negative coupling. The dimensionless parameters and variables are used below and the dimensionless time is also applied.

Let a starting point of system state be the stable trajectory with antiphase oscillations. The initial values of resistances $R_M = 0.8$, $R_m = 0.4$ are selected. The output signal $V_S = 0$ corresponds to the stable trajectories of initial state (**Figure 12**). The positive pulse with amplitude of 0.05 and duration of 0.1 arrives at time t = 0.9. It causes the delay of switching of MBO1. The series of four output pulses is generated (**Figure 12**) while the system is in an excited state and the antiphase is violated.

The difference Δ in the coupling factors significantly impacts on the speed of transition to a stable trajectory. If this value is small the transient process can be

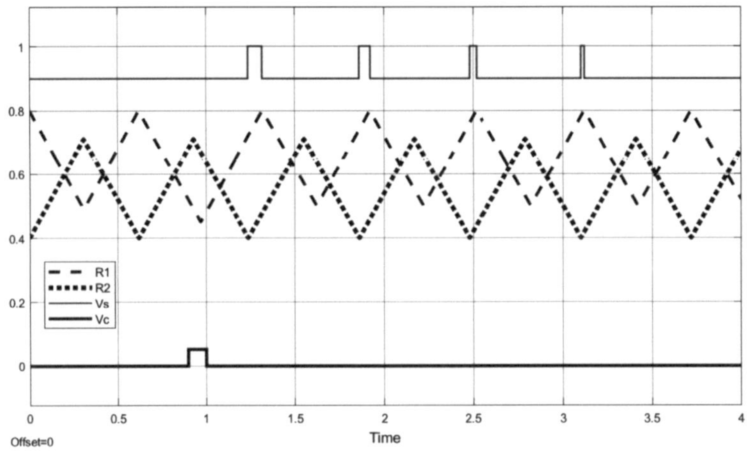

Figure 12.
Example of simulating the timing diagram of the generation of output pulse series in the coupled MBOs with inverting connections.

significantly delayed. At fixed values of the coupling strengths the input amplitude and time of arrival of the input pulse determine the time of transition to a stable trajectory. Due to this property the conversion of the input amplitude to the duration of the transition process can be performed.

4.3 Output

The system of two coupled MBOs with the inverting connection can be characterized by the following capabilities:

1. antiphase or in-phase steady state oscillations are generated depending on relation of coupling strengths;

2. the control signal causes transition to new stable state if the pulse amplitude is sufficient to change the threshold values;

3. appearance of pulses at the detector output is associated with the transition to new state and violation of the synchronization of oscillations;

4. the duration of the transient process and the number of pulses at the detector output are proportional to the amplitude of the drive signal and they are inversely proportional to the modulus of the difference in the coupling coefficients.

5. Properties of coupled memristor based oscillators for use in binary oscillator networks

The coupled MBOs have useful functional qualities for a number of applications. The possible application of connected MBOs as the basic elements of binary oscillation networks (BON) is discussed below. In particular, the BON with ring architecture and star-like architecture are considered. The presented before coupled MBOs with positive couplings and coupled MBOs with inverting connections are used for this purpose.

5.1 Features of binary oscillator networks based on memristor oscillators

The important properties of coupled MBOs are the simplicity of external managing the conditions of the oscillator injection locking, as well as fast frequency capture under relatively small impact amplitude.

The external excitation can violate the synchronicity of the coupled MBOs. The time to restore synchronization depends on the amplitude of the external impact and coupling strengths between the MBOs. As a result, the coupled MBOs provide the modulation of pulse trains desired for the implementation of oscillatory artificial neurons (AN).

By such a way, it becomes possible to apply the simpler coupling systems of the first order instead of using high-order nonlinear systems with reactive circuit elements and with high requirements for the element parameters.

Note that coupled MBOs belong to the class of binary oscillator and can be exploited in BON on base of integrated technologies. Using binary oscillators with binary output signals [16 18] is one of the promising lines for constructing oscillatory neural networks that are most suitable for integrated technologies. In such binary oscillator networks information is represented by binary streams.

The connection of N MBOs is described by system of N equations for variables $R_i(t)(i = 1, 2, \ldots, N)$. This system has the view of type Eq. (8):

$$\frac{dR_i(t)}{dt} = \begin{cases} -1, if\ R_i(t) > R_M - r_i(t), \\ 1, if\ R_i(t) < R_m - r_i(t), \\ \dfrac{dR_i(t-\Delta\tau, \Delta\tau \to 0)}{dt}, if\ R_M - r_i(t) > R_i(t) > R_m - r_i(t), \end{cases} \tag{15}$$

The outputs of the transmitting MBOs are connected to the inputs of the receiving MBOs directly or via logic gates. Therefore the variables $r_i(t)$ are binary functions of the outputs of the transmitting MBOs:

$$r_i(t) = k_i \cdot F_i\ (v_{out1}, v_{out2}, \ldots v_{outN}) \tag{16}$$

where $F_i(\ldots)$ is a logical function of N binary variables (0, 1). In this case the relationship between binary variable v_{out} and dR/dt is unambiguous.

The system of equations Eqs. (15) and (16) describes behavior of BON in autonomous mode. If there are external binary signals, they should be included into the F_i functions as additional external variables.

As simple examples of the elements of the binary oscillator networks based on memristor oscillators we can point out ring structure (**Figure 13**) which can be considered as extension of the considered before two coupled identical MBO with positive couplings (**Figure 5**) and also star-like structure (**Figure 14**) with applying the coupled MBO with possible inverting connections.

5.2 Some simulation results

Some simulation results to confirm the features of BON based on MBOs are presented below. To simulate the versions of BON fragments the dimensionless parameters of the variables were used: $R_M = 0.8$, $R_m = 0.4$.

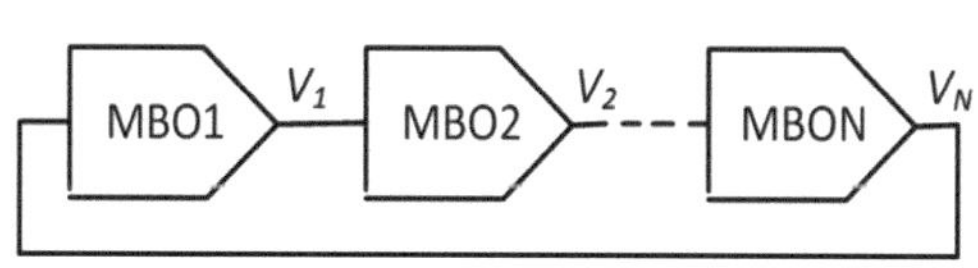

Figure 13.
Example of BON ring structure using memristor based oscillators.

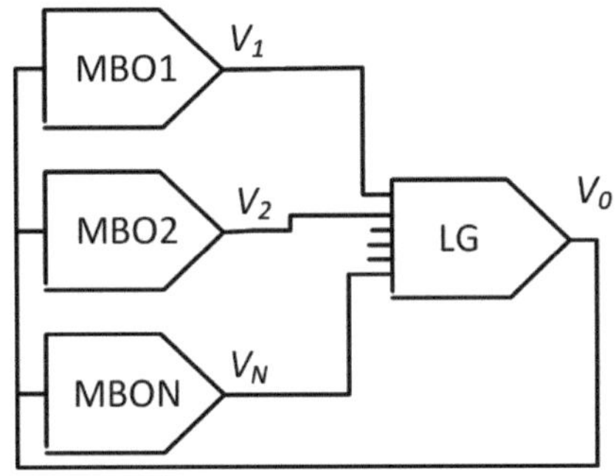

Figure 14.
Example of BON star-like structure using memristor based oscillators.

The circuit element of star-like structure containing two coupled MBOs was selected for simulation. This circuit contains also logical element OR. The coupling strength has the value r = 0.05.

The computed waveforms are given in **Figure 15**. The transient process from starting point to steady state is shown. The oscillators MBO1 and MBO2 have the different initial states: $R_1(0) = 0.4, R_2(0) = 0.8$. Then the oscillators tend to periodic steady state $R_1(t) = R_2(t)$ and reach it during four periods. The outputs of MBO1 and MBO2 are identical in steady state mode. This state corresponds to logical "0" due to applying the logical function XOR. The change of logical function XOR is shown in the lower curve in **Figure 15**. In this case the pulses of logical "1" appear under misalignment of the MBO1 and MBO2 outputs.

It can be mentioned that variation of initial state for MBO2 oscillator leads to change in duration of the process of steady state establishing (**Figure 15**). So for $R_2(0) = 0.7, 0.6, 0.5$ the transition to periodical steady state is performed for 3, 2 and 1 periods, respectively. In these cases, the difference in the initial states is a multiple of twice the coupling strength factor (here r = 0.05). This case corresponds to the complete synchronization of the oscillations (the phase shift is 0).

In connection with this non-multiple case is of interest. Then the initial states $R_1(0) = 0.43, R_2(0) = 0.8$ are chosen as examples. The corresponding computed waveforms are given in **Figure 16** for different versions.

As can be seen from **Figure 16a** the oscillations in MBO1 and MBO2 are synchronized with the shift at inverting absence. Steady state is reached for time interval of four periods, the periodic pulses are generated at the output, the phase detector XOR generates short mismatch pulses.

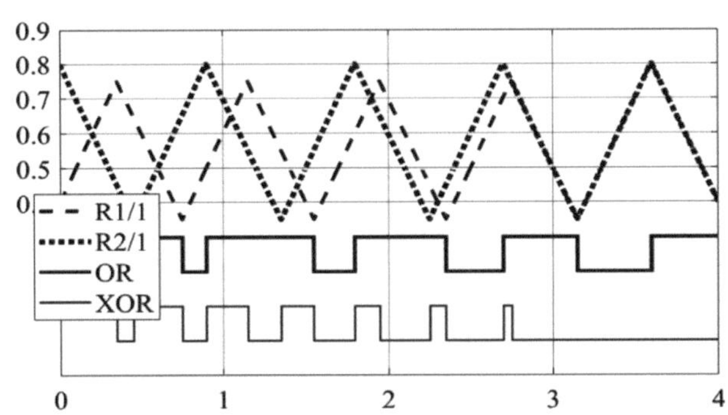

Figure 15.
The computed waveforms of the transient process from starting point to steady state for coupled MBOs: MBO1 - dashed line, MBO2 – dotted line, OR function -solid bold line, function XOR – solid line.

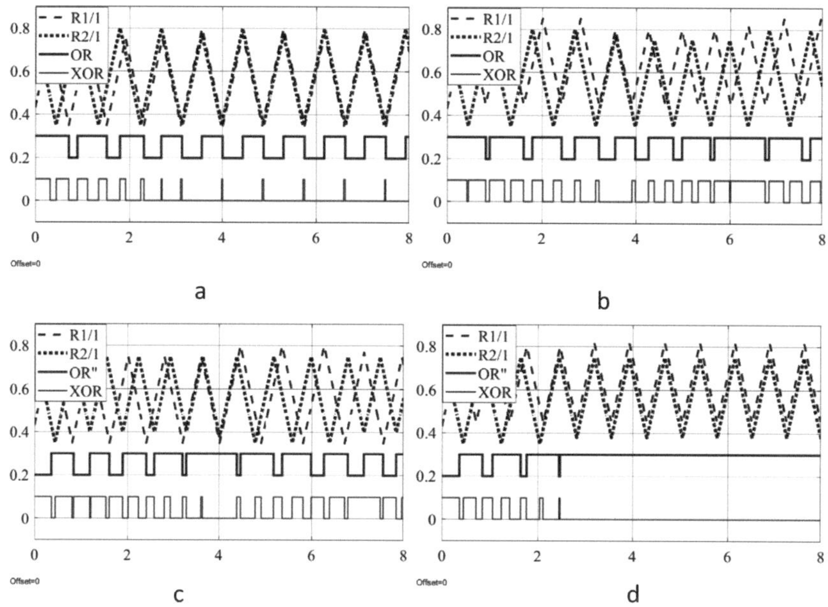

Figure 16.
The computed waveforms of the transient process for coupled MBOs with the initial states $R_1(0) = 0.43, R_2(0) = 0.8$. *The graphs correspond to the following versions of coupling types: (a) couplings without inverting; (b) couplings with analog inverting MBO1 input; (c) couplings with inverting of the MBO2 binary output; (d) couplings with joint inverting of both types.*

The timing diagram of MBOs behavior gets more complicated for cases with inverting the coupling signals (**Figure 16b**). The dependences of R (t) can be considered as modulated by triangular oscillations. The output signals of the logical element and the phase detector are converted into complex binary sequences with large period. The similar character of waveforms can be observed in the case of logical inversion (**Figure 16c**) when the logical OR circuit receives at the input the inverted signal from the output of MBO2. By such a way if the difference in the initial states is a non-multiple of twice the coupling strength factor then timing diagram of MBOs behavior is complicated for both types of inversion. It is interesting also that the simultaneous use of both types of inversion leads to another character of waveforms (**Figure 16d**). The full synchronization can be achieved after the transient process.

The presented simulation results illustrate the capabilities of coupled MBOs in its application as elements of Binary Oscillator Networks.

6. Conclusion

The chapter describes the behavior and application capabilities of the coupled reactance-less memristor based oscillators. This type of coupling memristor oscillators provides the generation of desired pulse trains with the complicate character of behavior. The chapter idea is to apply the simpler coupling systems of the first order instead of using high-order nonlinear systems with reactive circuit elements and with high requirements for the element parameters.

The coupled memristor-based reactance-less oscillators have the set of modes that is enough to provide the complex behavior desired in many applications.

To construct the memristor oscillator circuits the principle of controlling threshold parameters is applied. The constructing the piecewise constant memristor oscillators is one of advantages of this approach.

Two types of oscillator couplings are analyzed in chapter: coupled memristor based oscillators with positive couplings, coupled memristor based oscillators with inverting connections.

The coupling memristor based oscillators can be considered as the analog-to-digital converters that provide transform of amplitude variation. The properties of coupled reactance-less memristor based oscillators open up the possibility of constructing binary oscillator networks on its base for solving a wide range of problems. In particular, star-like binary oscillation networks based on coupled memristor oscillators with only one logical elements create a number of promising applications, including oscillator reservoir calculations, stochastic oscillators, neural networks with probabilistic coding.

Acknowledgements

The reported study was funded by RFBR, project number 19-29-03012.

Conflict of interest

The authors declare no conflict of interest.

Author details

Vladimir V. Rakitin and Sergey G. Rusakov*
Institute for Design Problems in Microelectronics of Russian Academy of Sciences (IPPM RAS), Moscow, Russia

*Address all correspondence to: rusakov@ippm.ru

References

[1] Chua L.O., Memristor-missing circuit element. IEEE Trans. Circuit Theory. 1971, V. 18, N. 9, pp. 507–519.

[2] Chua L.O., Sirakoulis G. Ch. A. Adamatzky A., editors. Handbook of Memristor Networks. 2019. 1368 p. h ttps://doi.org/10.1007/978-3-319-76375-0_1

[3] Strukov DB, Snider GS, Stewart DR, Williams RS. The missing memristor found. Nature. 2008;453(7191):80-83

[4] Schuman C., Potok T., Patton R., Birdwell J., Dean M., Rose G., Plank J. A Survey of Neuromorphic Computing and Neural Networks in Hardware. arXiv:1705.06963. 2017. V1, 19.

[5] Islam R., Li H., Chen P., Wan W., Chen H., Gao B, Wu H., Yu S., Saraswat K., Wong H. Device and Materials Requirements for Neuromorphic Computing. Journal of Physics D: Applied Physics. 2019. V.52. N.11.

[6] Hopfield J.J. Neural networks and physical systems with emergent collective computational abilities. Proc. Natl. Acad. Sci. USA. 1982. 79(8). pp.2554–2558.

[7] Tamas R, Chua L.O. The CNN universal machine: an analogic array computer. IEEE Transactions on Circuits and Systems II: Analog and Digital Signal Processing. 1993. V. 40, N. 3. pp.163-173.

[8] Izhikevich E. M. Dynamical systems in neuroscience. MIT press. 2007.

[9] James A., editor. Memristor and Memristive Neural Networks, IntechOpen, 2018. 324 p DOI:10.5772/66539

[10] Itoh M., Chua L.O. Memristor Oscillators. Int. J. Bifurcation Chaos. 2008. V. 18. N. 11. pp. 3183–3206.

[11] Talukdar A., Radwan A., Salama K. Generalized Model for Memristor Based Wien Family Oscillators. Microelectronics Journal. 2011. V. 42. N. 9.pp. 1032–1038.

[12] Talukdar A., Radwan A., Salama K. Non Linear Dynamics of Memristor Based 3 Order Oscillatory System. Microelectronics journal. 2012. V 43. N.3.pp. 169–175.

[13] Wang Y., Liao X. Stability analysis of multimode oscillations in three coupled memristor-based circuits. AE ËU -International Journal of Electronics and Communications. 2016. V.70, N. 12, pp. 1569–1579.

[14] Banchuin R. On the fractional domain generalization of memristive parametric oscillators. Cogent Engineering. 2019. 6:1, 1617094, DOI: 10.1080/23311916.2019.1617094

[15] Elsonbaty A., Abdelkhalek A., Elsaid A. Dynamical behaviors of coupled memristor-based oscillators with identical and different nonlinearities. Mathematical Problems in Engineering. 2018. pp.1-19.

[16] Forti M., Corinto F. Complex Dynamics in Arrays of Memristor Oscillators via the Flux-Charge Method. IEEE Transactions on Circuits and Systems I: Regular Papers. 2017.V.66. N.11. pp.1-11.

[17] Ponce E., Ros J., Freire E., Amador A. Unravelling the dynamical richness of 3D canonical memristor oscillators. Microelectronic Engineering. 2017. pp.1-22.

[18] Saha, D.C., Saha, P., Ray, A., Roychowdhury, A., On the Synchronization of Synaptically Coupled Nonlinear Oscillators: Theory and Experiment. Annual Review of Chaos Theory, Bifurcations and Dynamical Systems. 2016. V.6. pp.1-29.

[19] Fouda M., Radwan A. Memristor-based voltage-controlled relaxation oscillators. Int. J. Circ. Theor. Appl. 2013. V. 42, N. 10, pp.1092-1102.

[20] Zidan M., Omran H., Smith C., Radwan A. Salama K. A Family of Memristor Based Reactance-Less Oscillators. Int. J. Circuit Theory and Applications. 2013. V. 42. № 11. pp. 1103–1122.

[21] Kyriakides E, Georgiou J. A compact, low-frequency, memristor-based oscillator. Int. J. Circ. Theor. Appl. 2015. V.43 pp.1801-1806.

[22] Radwan A., Fouda M. On the Mathematical Modeling of Memristor, Memcapacitor, and Meminductor. Cham: Springer International Publishing. Switzerland. 2015.

[23] Khatib M., Mosad, A., Fouda, M., Radwan, A. Generalized Analysis of Symmetric and Asymmetric Memristive Two-Gate Relaxation Oscillators. IEEE Trans. Circuits Syst. I, 2013. V. 60. N. 10. pp. 2701-2708.

[24] El-Naggar A., Fouda M., Madian A., Radwan A. Reactance-less RM relaxation oscillator using exponential memristor model. 2016. 28th International Conference on Microelectronics (ICM). Giza, pp. 361-364, doi: 10.1109/ ICM.2016.7847890.

[25] Rakitin V., Rusakov S. The Signal Converter Based on Monostable Memristor Oscillator. Problemi Razrabotki Perspektivnih Mikro- i Nanoelektronnih system (MES). 2017 (1). pp. 29-32

[26] Rakitin V., Rusakov S. Operating principles of reactance-less memristor-based oscillators. Journal of Communications Technology and Electronics. 2017. V. 62, N.6, pp.621–625.

[27] Rakitin V., Rusakov S. Principles of the Functioning of Nonreactive Double Memristor Oscillators. Journal of Communications Technology and Electronics. 2019. V.64, N.6, pp.622–628.

[28] Rakitin V., Rusakov S. The Reactance-Less Two-Memristor based Oscillator for Signal Processing. In: 2020 Moscow Workshop on Electronic and Networking Technologies (MWENT). 2020. pp.1-5.

[29] Mazumder P., YilmazY., Ebong I., Lee W. Memristor-based Cellular Nonlinear/Neural Network: Design, Analysis and Applications. Neuromorphic Circuits for Nanoscale Devices. 2018. pp.275-302.

[30] Mazumder P., Yilmaz Y., Ebong I., Lee, W. Dynamic Analysis of Memristor-based Neural Network and its Application. Neuromorphic Circuits for Nanoscale Devices. 2018. pp.303-350.

[31] Xu B., Lin H., Wang G. Hidden. Multistability in a Memristor-Based Cellular Neural Network. Advances in Mathematical Physics. 2020 V.2020, no.9708649, pp.1-10.

[32] Majdabadi M. Shamsi J., Shokouhi S. Hybrid CMOS/memristor crossbar structure for implementing Hopfield neural network. Analog Integrated Circuits and Signal Processing. 2020. pp.1-13.

[33] Nikonov D., Csaba G., Porod W., Shibata T., Voils D., Hammerstrom D., Young I. Bourianoff G. Coupled-Oscillator Associative Memory Array Operation for Pattern Recognition. IEEE Journal on Exploratory Solid-State Computation Devices and Circuits. 2015. V. 1, pp. 85-93.

[34] Csaba G., Porod W. Coupled oscillators for computing: A review and perspective. 2020. Appl. Phys. Rev. 7, 011302 (2020); doi: 10.1063/1.5120412

[35] Raychowdhury A., Parihar A., Smith G. Narayanan V., Csaba G.,

Jerry M., Porod W.Datta S. Computing with Networks of Oscillatory Dynamical Systems. Proceedings of the IEEE 2019. V 107, pp.73–89.

[36] Hsieh C. C., Chang Yao-Feng, Chen Y.-C. et al., Review of Recently Progress on Neural Electronics and Memcomputing Applications in Intrinsic SiOx-Based Resistive Switching Memory. Book Chapter: "Memristor and Memristive Neural Networks", ISBN: 978-953-51-5481-5. InTech. 2018. http://dx.doi.org/10.5772/intechopen.68530

[37] Wang Y., Wang G., Shen, Y., Iu H. A. Memristor Neural Network Using Synaptic Plasticity and Its Associative Memory. Circuits, Systems, and Signal Processing. 2020. pp.1-16.

[38] Luo B., Li Y., Liu D., Yang Z., Zhu, Y. Adaptive synchronization of memristor-based neural networks with discontinuous activations. Neurocomputing. 13th November 2019. pp.1-31

[39] Yang L., Zeng Z., Shi X. A memristor-based neural network circuit with synchronous weight adjustment. Neurocomputing. 18th July 2019. pp.1-11.

[40] Chua L.O., Kim, H., Sah, M., Roska, T., Yang, C. Memristor Bridge Synapses. Proceedings of the IEEE. 2011. V.100, N.6, pp. 2061-2070.

[41] Secco J., Poggio M., Corinto F. Supervised neural networks with memristor binary synapses. International Journal of Circuit Theory and Applications. 2018. V.46. N.1. pp.1-13.

[42] Arthur J., Boahen K. Silicon-Neuron Design: A Dynamical Systems Approach. IEEE Transactions on Circuits and Systems I: Regular Papers. 2011. V. 58, pp. 1034- 1043. DOI: 10.1109/TCSI.2010.2089556.

[43] Andrew S., Cassidy A., Merolla P., Arthur J. et all. Cognitive Computing Building Block: A Versatile and Efficient Digital Neuron Model for Neurosynaptic Cores. International Joint Conference on Neural Networks. 2013. pp.1-10. DOI:10.1109/IJCNN.2013.6707077

[44] Tsubone T., Saito T. Manifold piecewise constant systems and chaos. IEICE Trans. Fundamentals. 1999. E82-A, N 8, pp.1619-1626.

[45] Matsuda C., Torikai H. A Novel Generalized PWC Neuron Model: Theoretical Analyses and Efficient Design of Bifurcation Mechanisms of Bursting. IEEE Transactions On Circuits and Systems II: Express Briefs. 2012. V. 11, N. 4.

[46] Yamashita Y., Torikai H., Theoretical Analysis for Efficient Design of a Piecewise Constant Spiking Neuron Model. IEEE Transactions On Circuits and Systems II: Express Briefs. 2014. V. 61, N. 1, pp 54-58.

[47] Mitsubori K., Saito T. Dependent Switched Capacitor Chaos Generator and Its Synchronization. IEEE Transactions On Circuits and Systems. I. 1997. V. 44, N. 12, pp. 1122–1128.

[48] Tsubone T., Saito T., Inaba N. Design of an analog chaos-generating circuit using piecewise-constant dynamics. Prog. Theor. Exp. Phys., 053A01, 2016.

[49] Rakitin.V., Rusakov S. Coupled Piecewise Constant Memristor based Reactance-less Oscillators. IEEE East-West Design & Test Symposium (EWDTS), 2020.

[50] Rakitin.V., Rusakov S. Memristor based Oscillators with Controlled Threshold Parameters. European Conference on Circuit Theory and Design (ECCTD), 2020.

[51] Yang J., Strukov D., Stewart D. Memristive devices for computing.

Nature Nanotechnology.2013. V.8. N.1. pp.13-24.

[52] Vourkas I., Sirakoulis G. Memristor-Based Nanoelectronic Computing Circuits and Architectures. Springer 2016 DOI 10.1007/978-3-319-22647-7.

[53] Burr G., Shelby R., Sebastian A., et al. Neuromorphic computing using non-volatile memory. Advances in Physics: X. 2017;2(1):89-124

[54] Suri M., editor. Advances in Neuromorphic Hardware Exploiting Emerging Nanoscale Devices. Springer. 2017. 210 p. DOI 10.1007/978-81-322-3703-7

[55] Liang H., Cheng H., Wei J., Zhang L., Yang L., Zhao Y., Guo H. Memristive Neural Networks: A Neuromorphic Paradigm for Extreme Learning Machine. IEEE Transactions on Emerging Topics in Computational Intelligence. 2019. V. 3, N.1, pp.15-23.

[56] Wang W. Binary-oscillator Networks: Bridging a Gap Between Experimental and Abstract Modeling of Neural Networks. Neural Comput. 1996. N 8. pp. 319-39.

[57] Joao V., Giacomin E., Qureshi Y., Zapater M., Tang, X, Kvatinsky, S., Atienza, D., Gaillardon P.-E. A Product Engine for Energy-Efficient Execution of Binary Neural Networks Using Resistive Memories. In: IFIP/IEEE International Conference on Very Large Scale Integration (VLSI-SoC). 2019. pp.1-6

[58] Pham K., Nguyen T., Tran S., et. al. Memristor Binarized Neural Networks. Journal of Semiconductor Technology and Science (JSTS). 2018. V.18. N.5. pp.568-577

[59] Elwakil A. S. and Ozoguz S., "A low frequency oscillator structure," 2009 European Conference on Circuit Theory and Design, Antalya, Turkey, 2009, pp. 588-590, doi: 10.1109/ECCTD.2009.5275054.

[60] Zidan MA, Omran H, Radwan AG, Salama KN. Memristor-based reactance –less oscillator. Electronics Letters 2011; 47, pp. 1220–1221.

[61] Azghadi R., Chen Y.-C., Eshraghian J. K., Chen, J., Lin C.-Y., Amirsoleimani, A., Chang Yao-Feng et al. Complementary metal-oxide semiconductor and memristive hardware for neuromorphic computing. Adv. Intell. Syst. 2:1900189. 2020. pp. 1-24. doi: 10.1002/aisy.201900189

Chapter 8

Practical Approach to Induce Analog Switching Behavior in Memristive Devices: Digital-to-Analog Transformation

Firman Mangasa Simanjuntak, Sridhar Chandrasekaran, Debashis Panda, Aftab Saleem and Themis Prodromakis

Abstract

The capability of memristor devices to perform weight changes upon electrical pulses mimics the analogous firing mechanism in biological synapses. This capability delivers the potential for neuromorphic computing and pushes renewed interests in fabricating memristor with analog characteristics. Nevertheless, memristors could often exhibit digital switching, either during the set, reset, or both processes that degenerate their synaptic capability, and nanodevice engineers struggle to redesign the device to achieved analog switching. This chapter overviews some important techniques to transform the switching characteristics from digital to analog in valence change and electrochemical metallization types memristors. We cover physical dynamics involving interfacial diffusion, interfacial layer, barrier layer, deposition, and electrode engineering that can induce digital to analog switching transformation in memristor devices.

Keywords: digital, analog, switching transformation, switching mode, resistive memory, synaptic plasticity

1. Introduction

Memristive devices offer a great promise not only for ultra-high density data storage [1] but also for in-memory computing applications [2]. In-memory computing utilizes the synaptic plasticity of the memristors, where they can mimic the biological synapse or neuron (neuromorphic) [3]. A lot of parameter should be considered to achieve high-performance memristor-based artificial synapses (dynamic range, linearity, asymmetry, level of states, etc.) [4]. Much effort has been conducted to improve these parameter, such bi-layering [5], irradiation [6], doping [7], and deposition engineering [8]. Most of synaptic parameters can also be enhanced by tuning the potentiation and depression pulse schemes [5]. Nevertheless, none of these parameters can be performed by memristors that exhibit digital characteristics. On the other hand, several challenges exist in fabricating analog memristors, and, most often, the fabricated memristors exhibit semi- or pure digital characteristics. In this chapter, we review some important methods to induce analog characteristics by nanofabrication and electrical engineering that could provide useful insight for the device development engineers to transform their digital memristors into analog.

Resistance switching in memristor devices, set (HRS-to-LRS) and reset (LRS-to-HRS) processes, can occur either in abrupt or gradual resistance change. **Figure 1** depicts the schematic of current–voltage and conductance-pulses curves of the abrupt and gradual resistance changes [2]. The abrupt resistance change is a phenomenon where the current or conductance of the resistance states is suddenly changed at a threshold voltage, as depicted in **Figure 1(a)** and **(b)**. Any memristor devices with this digital characteristic tend to have limited capability to exhibit multiple resistance states (multibit performance). Henceforth, this abrupt behavior is also called digital switching. The only possible way to induce multibit performance is by varying the current compliance level to limit the size of the conduction filaments, and thus it controls the amount of the current that can pass through the cell (**Figure 1(b)**). On the other hand, gradual resistance change is a phenomenon where the device does not require any threshold voltage to change the current or conduction of the states, as depicted in **Figure 1(c)** and **(d)**. In this case, any given voltage or electrical pulse stimulus is able to modulate the current or conduction of the states [9]; thus, the number of the states that the memristor can exhibit is equivalent to the number of voltage or pulse stimulus it can response to (**Figure 1(d)**). Henceforth, this gradual behavior is also called analog switching.

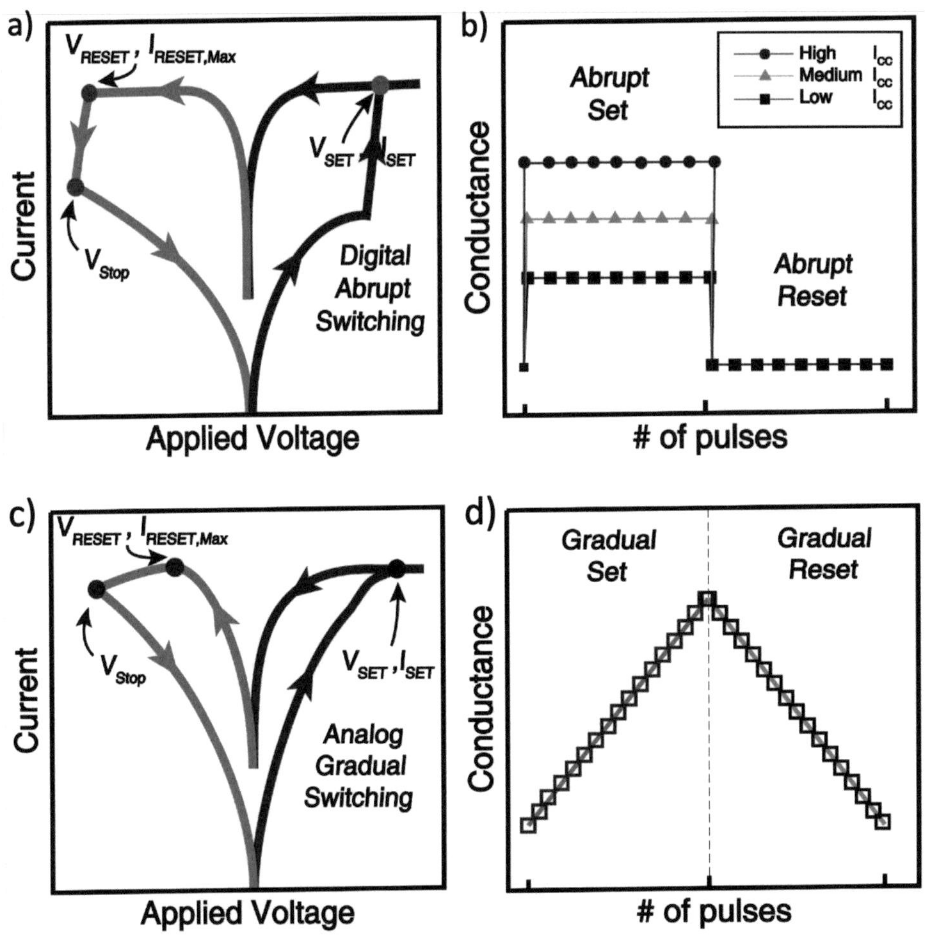

Figure 1.
Schematic of (a, b) digital and (c, d) analog switching behaviors. Reprinted from [2].

It is important to note that, based on our knowledge, the memristor, which works under unipolar mode (employing the same voltage polarity to set and reset the device), cannot show analog switching due to the rapid Joule heating process. Hence, the memristor should be designed in such a way where the Joule heating should not play a major role in its switching mechanism. Bipolar mode, however, has several variants of mode, such as conventional, complementary, and diode-like bipolar modes. **Figure 2** shows the conventional and complementary bipolar resistive switching in Pt/ZnO/TiN memristor that was observed by Khan SA et al. [10]. They suggested that the two modes can be exhibited by varying the bias condition; counter-clockwise and clockwise voltage sweeps exhibit conventional (BRS) and complementary (CRS) bipolar modes, respectively, as depicted in **Figure 1(a)** and **(b)**. The efficacy of the modes on

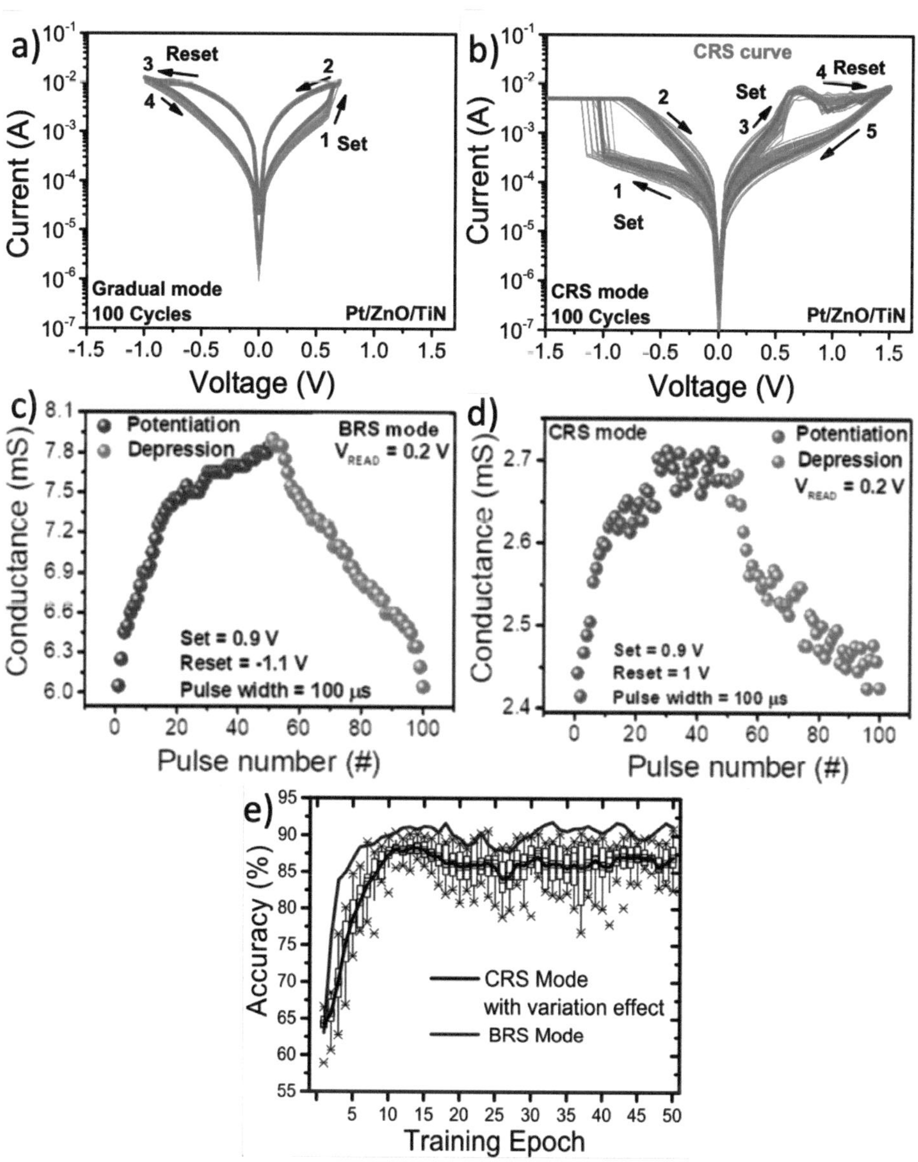

Figure 2.
I-V curves of (a) conventional analog bipolar (BRS) and (b) complementary (CRS) resistance switching modes in Pt/ZnO/TiN memristive device. Synaptic potentiation and depression of (c) BRS and (d) CRS. Epoch training accuracy of BRS and CRS devices. (e) Training accuracy of CRS and BRS modes. Reprinted from [10].

the synaptic performance was studied **Figure 1(c)** and **(d)**. Although both modes can exhibit synaptic plasticity (potentiation and depression characteristics), BRS performs superior linearity than that of CRS and, thus, better training accuracy (**Figure 2(e)**).

Memristors having a complementary or a diode-like mode offer benefit to the circuit integrations, which they do not need to be stacked with an additional select device in the array configuration [11, 12]. However, the exact switching mechanism of the complementary and diode-like bipolar modes is still not fully understood. Likewise, the synaptic properties of the complementary and diode-like memristor devices are still less investigated. Henceforth, based on these reasons, the present chapter only focuses on the device development on the conventional bipolar memristive devices.

2. Digital-to-analog switching transformation

In some cases, a memristor device can show both abrupt and digital behaviors in the same switching cycle, such as digital set and analog reset or vice-versa. This behavior may induce better multibit performance, but it may not be sufficient to induce satisfactory synaptic performance. For example, if the set or reset process exhibits digital behavior then we can assume that the device will not exhibit good potentiation or depression, respectively. Henceforth, it is important to have analog switching for both set and reset processes. We discuss several important techniques to induce analog behavior in memristor devices. These techniques include transforming the switching conduction from filamentary to homogeneous, adding an interfacial layer below the top electrode, and electrode engineering.

2.1 Filamentary to homogeneous switching transformation

Despite the filamentary switching is more scalable friendly, as the localized filaments (approximately 20 nm) [13] take part in the switching process than the homogeneous switching, which relies on the conduction changes of the entire bulk structure. However, the homogeneous switching tends to be easier to exhibit analog behavior. The co-existence of both filamentary and homogeneous resistive switching in a single device can be observed by adjusting the electrical operation (biasing condition).

Huang C-H et al. [14] reported that a digital unipolar in Pt/ZnO/Pt device could be controlled to show analog bipolar switching mode after reversing the bias condition at higher current compliance (CC), as depicted in **Figure 3(a)**. The reversed bias made the switching layer consists of two regions, the oxygen-rich region (below the top electrode) and the oxygen-deficient region (above the electrode), which transformed the filamentary into homogeneous switching. The reversed-biased technique can also be useful to transform filamentary bipolar to homogeneous bipolar, as reported by Ryu H. and Kim S. shown in **Figure 3(b)** [15]. They observed that applying appropriate stop voltage sweep prior to the LRS, at the negative differential region (NDR) voltage regime, can induce homogeneous switching in the Pt/Al_2O_3/TiN devices. Even though the *I-V* hysteresis of homogeneous switching is less obvious than the filamentary, it was reported that the synaptic behavior of the homogeneous switching is significantly improved, confirming the analog nature of the device [15]. A similar result was also observed in Pt/WO_x/W device [16].

The homogeneous switching tends to dominate when the device operates at a lower current regime. Li Y. et al. [17] reported that the homogeneous switching in Ag/NiO/Pt device can be observed prior to the electroformed process, as depicted

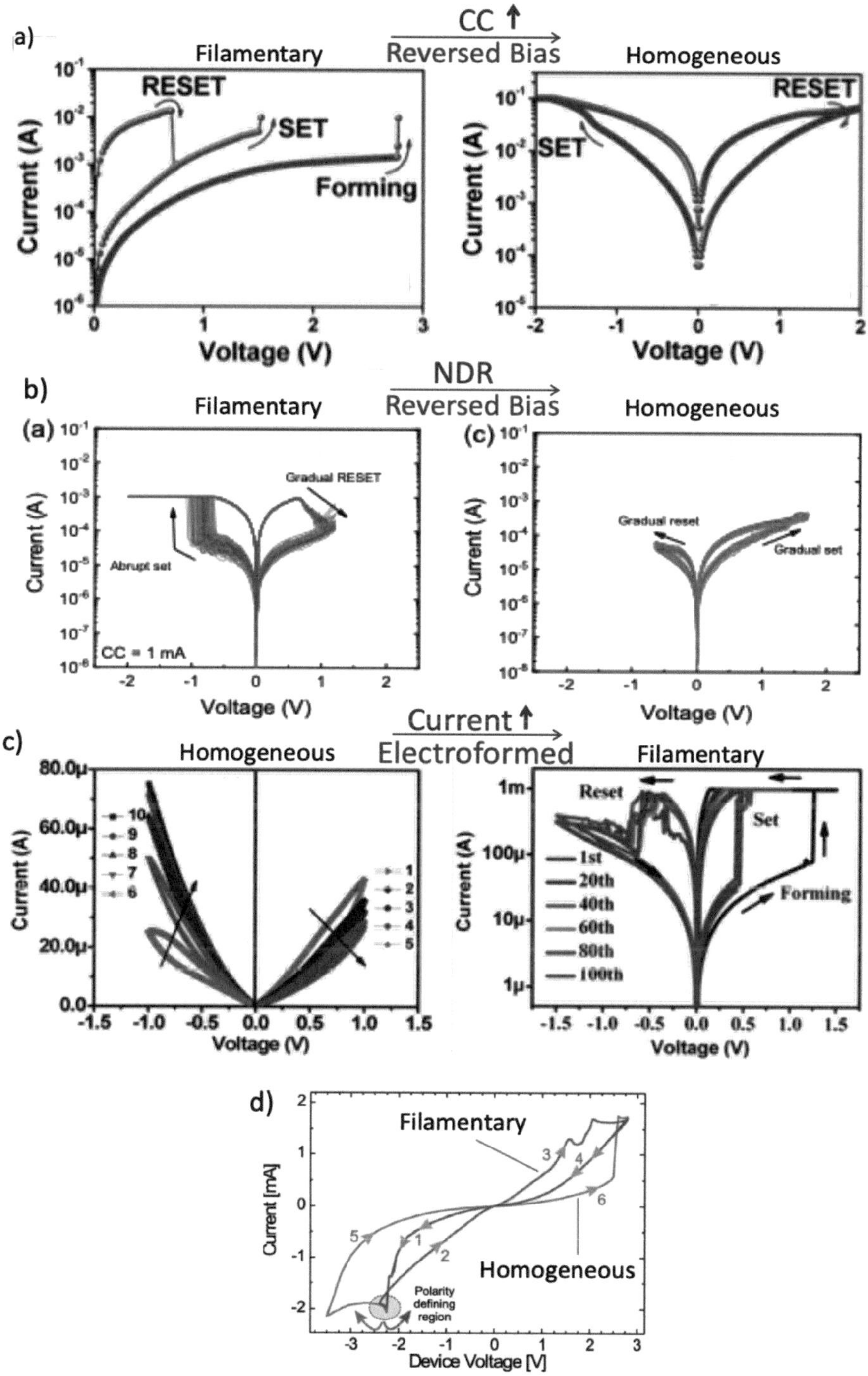

Figure 3.
Several strategies to induce homogeneous switching by controlling the biasing condition. The employment of (a) reversed bias and higher CC in Pt/ZnO/Pt device [14], (b) reversed bias at the NDR region in Pt/Al_2O_3/TiN device, [15] (c) successive voltage sweeps prior to the electroformed in the Ag/NiO/Pt device [17], and (d) opposite polarity in Fe-doped $SrTiO_3$ device [19]. Reprinted and adapted from [14, 15, 17, 19].

in **Figure 3(c)**. The successive voltage sweeps during positive and negative voltage incessantly decrease and increases the memristor conductance, respectively. Operating the device at a lower current regime unable to form the filament, but it is sufficient to control the movement of the intrinsic defects within the bulk and, thus, modulate the Schottky barrier height (SBH). A variation of SBH was observed in this device at different scans, due to the widened of the Ag/p-NiO interface depletion width. A possible conduction mechanism of filament formation was also conferred there. Interestingly, as discussed there, the low temperature treated NiO device is not showing analog switching. A similar analog switching result was shown in PT/BiFeO3/Pt device [14]. The Ag/CuAlO$_2$/TiO2/p^{++}-Si structure was also shows similar analog switching due to the Ag ions and oxygen migration under the electric field [18]. In some cases, homogeneous and filamentary switching can also co-exist at the same current regime as well. Muenstermann R. et al. reported that the Pt/SrTiO$_3$(Fe)/Nb:SrTiO$_3$ device exhibited non-polar behavior, as shown in **Figure 3(d)** [19]. Intriguingly, the counter eightwise and eightwise switching are controlled by different switching mechanism which is filamentary and homogeneous switching, respectively.

Kim S. et al. [20] reported that the analog switching can also be achieved by a partial reset scheme. They observed that the reset process of the Cu/HfAlOx/Si device consist of two stages where the first stage (partial reset) is controlled by an electric field and the second stage (full reset) is dominated by Joule heating mechanism, as depicted in **Figure 4(a)**. **Figure 4(b)** and **(c)** show the device exhibits digital or analog switching when it operates with a full reset or a partial reset, respectively. As expected, device that was operated with a partial reset performs better potentiation and depression than that of the full reset one, as shown in **Figure 4(d)** and **(e)**. It is still not clear the physics behind this phenomenon; however, we hypothesis that this may due to the filamentary to homogenous switching transformation as well. Nevertheless, the relationship between the conduction mechanism, analog switching, and synaptic behavior should be investigated further.

2.2 Insertion layer engineering

Analog switching can be induced by inserting a metal film between the electrode and the storage layer. Here, we discuss insertion layer techniques that can transform

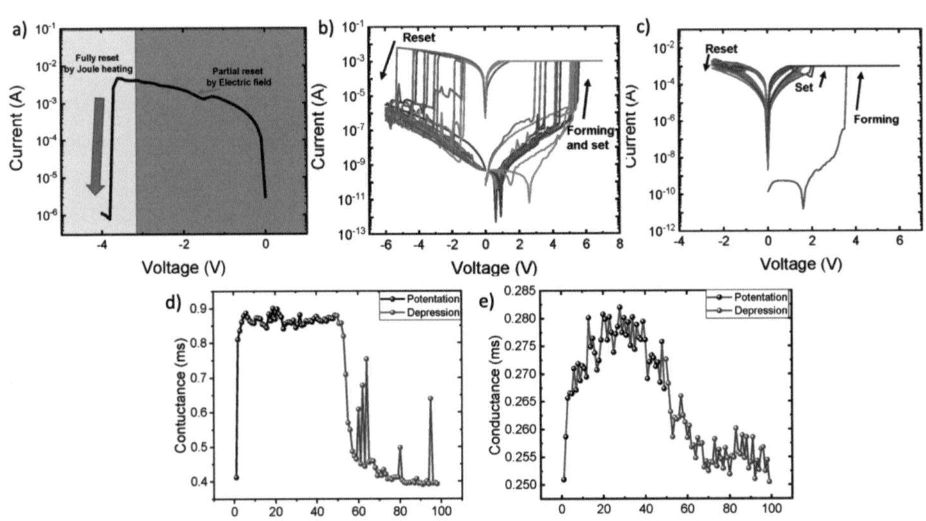

Figure 4.
(a) Two stages of reset process in Cu/AlHfOx/Si device. Switching characteristics of devices having (b) full and (c) partial resets. Synaptic behavior of the devices that operate with (d) full and (e) partial resets. Reprinted and adapted from [20].

valence change and conductive-bridge type memristors from digital into analog switching. In these techniques, a metal layer is inserted between the top electrode and storage layer to control the drift of the anions and cations defects during the switching process.

2.2.1 Oxygen scavenging layer

The formation and rupture of the oxygen conducting filament in the valence change memristor are controlled by redox of oxygen, where it is mainly taken place at the electrode/oxide interfaces [6]. Hence, in order to achieve analog behavior, we need to ensure that the oxygen ions that injected to- (set process) or from (reset process) the electrode should be continuously drifted during the entire switching process.

Chang L-Y et al. suggest that the continuous drift can be done by inserting a metal layer that has similar Gibbs free energy of oxide formation (ΔG_f) value to the storage layer [21]. **Figure 5(a)** and **(b)** show the *I-V* curves of TiN/TiO_2/Ti and TiN/Ti/TiO_2/TiN devices, respectively; the thickness of the Ti insertion layer was 4 nm. It is observed that the device without Ti insertion layer (0Ti) exhibits digital switching; conversely, analog switching can be observed after the insertion layer was employed (4Ti). This digital-to-analog switching transformation is further confirmed by the behavior of the synaptic plasticity of the devices, as shown in **Figure 5(c)** and **(d)**. Under a given pulse scheme, the conductance change of the 0Ti device rises (potentiation) and falls (depression) abruptly; meanwhile, the synaptic plasticity in the 4Ti device is more gradual.

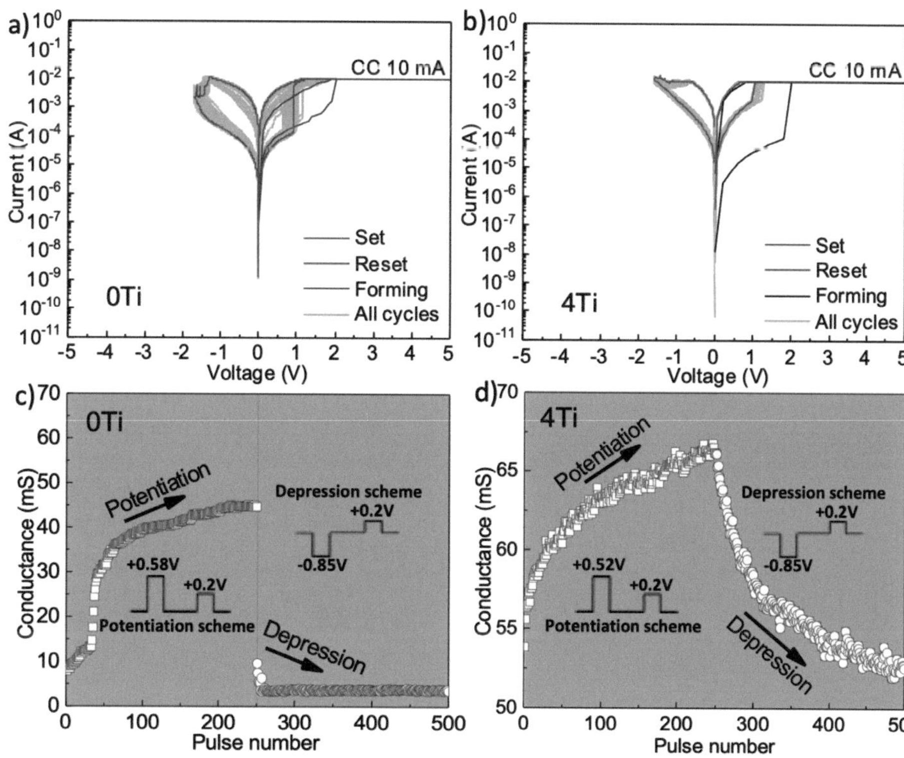

Figure 5.
Typical I-V curves of devices made (a) without (0Ti) and (b) with Ti (4Ti) insertion layer. Potentiation and depression synaptic plasticity of (c) 0Ti and (d) 4Ti devices. Reprinted from [21].

Chang L-Y et al. compared the interfacial properties between the stacks made without and with Ti layer, as depicted in **Figure 6**; for this purpose, they inserted 20 nm thick of Ti (20Ti) to obtain a more obvious reaction at the interface. Based on the depth-XPS analysis (**Figure 6(c–e)**), the Ti layer absorbed oxygen from the TiO_2 layer and forming TiOx interfacial layer at the TiN/TiO_2 interface. Note that the formation of the interfacial layer was occurred during the deposition process (pristine stack). Based on the material analysis, they proposed that TiOx interfacial layer can gradually ionize the oxygen ions during set/potentiation and reset/depression process that promotes the occurrence of gradual switching in the device (**Figure 6(f)**).

2.2.2 Cation drift barrier layer

The conductive-bridge type memristor utilizes the electrochemical metallization process where the atoms from the active electrode (Cu, Ag, or Ni) drift to the switching layer and form the conduction bridge (filament) [22, 23]. A diffusion barrier is usually employed in this type of memristor device to control the atomic drift [24, 25]. Aftab S. et al. used an oxidizable metal TiW as a diffusion barrier layer inserted between the switching layer and active metal top electrode to transform TaOx-based memritor from digital set into analog [26]. **Figure 7** shows the I-V curves and synaptic behavior of Cu/TaOx/TiN (device A) and Cu/TiW/TaOx/TiN (device B) memristors. It was observed that the device made without the TiW barrier layer exhibits digital set with a poor synaptic behavior.

In **Figure 7(a)** and **(b)** I-V curves for both devices is shown. The forming compliance current for both devices is 500 μA and set/reset cycle compliance current is 1 mA. The device with 20 nm TiW barrier layer (Device B) insertion clearly shows gradual switching for both set and reset cycles as compared to without barrier layer (Device A) device. The gradual behavior superiority is further confirmed by synaptic plasticity as shown in **Figure 7(c)** and **(d)**. By using an optimized pulse scheme with pulse width of 10 μs they observed abrupt conductance change when positive pulses are applied for potentiation process in Device A. However, Device B shows gradual change in conductance states when positive pulses are applied for

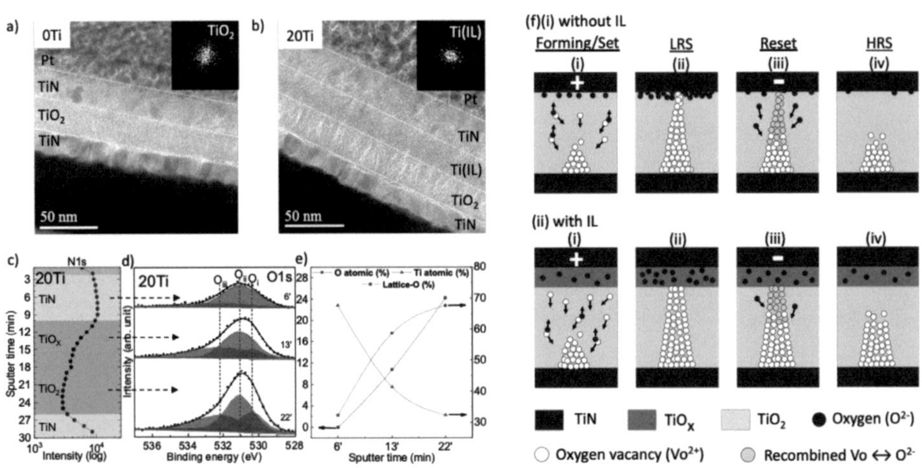

Figure 6.
Cross-sectional TEM image of (a) 0Ti and (b) 20Ti device stacks. (c) Intensity of N1s core level and (d) deconvolution of O1s core level spectra at various depth within the 20Ti stack. (e) Concentration of oxygen and titanium elements, and lattice-oxygen in the respective depth. (f) Conduction mechanism of the devices. Reprinted from [21].

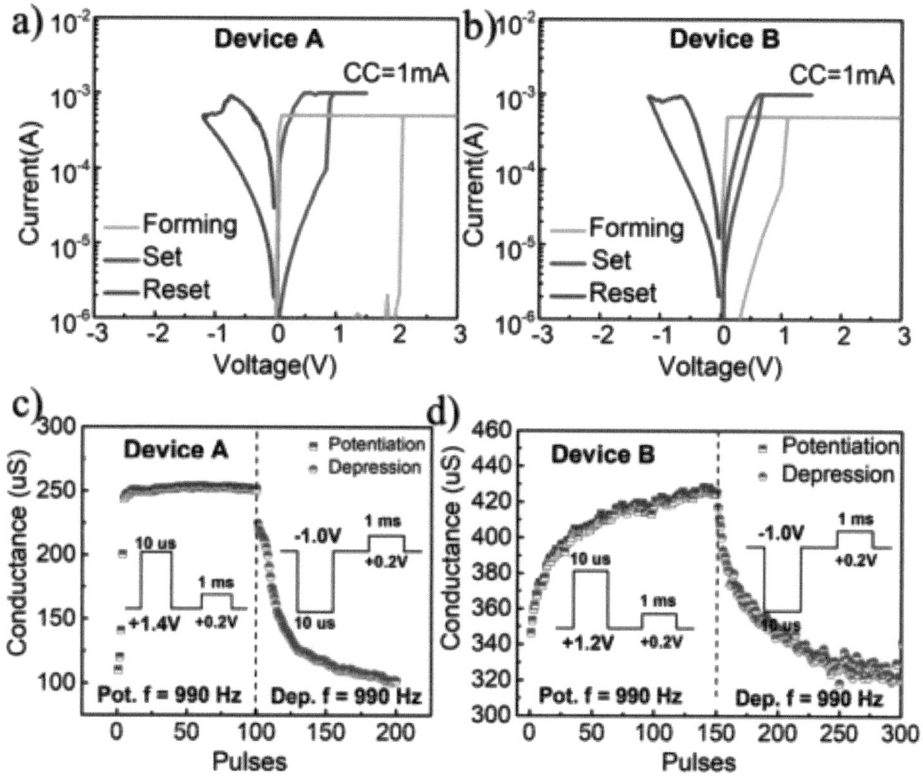

Figure 7.
Typical I-V curves and synaptic behavior of devices made without (device A) and with (device B) TiW barrier layer. Reprinted from [26].

potentiation. Note that they also observed a significant improvement in data retention of the device made with a barrier layer.

The inserted barrier layer restricts excessive metal ions diffusion into the TaOx layer and forms an oxygen vacancy-rich TiWOx layer at the interface, as depicted in **Figure 8(a–d)**. After the insertion of barrier layer, Cu diffusion into

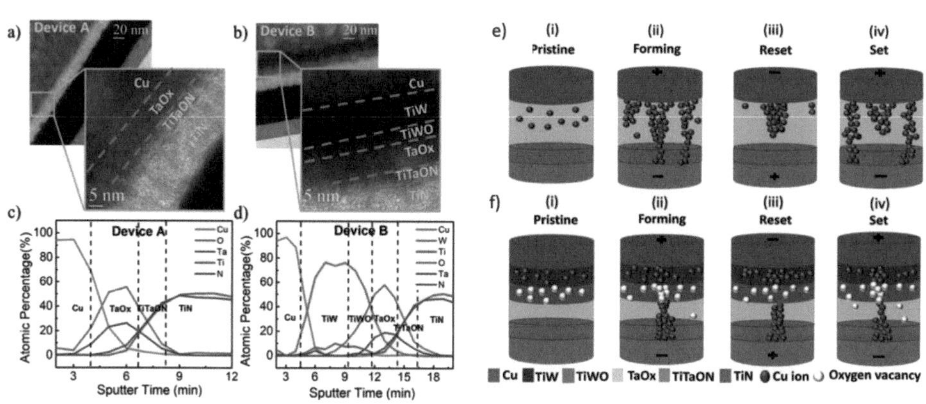

Figure 8.
Cross-sectional TEM image of (a) 0Ti and (b) 20Ti device stacks. (c) Intensity of N1s core level and (d) deconvolution of O1s core level spectra at various depth within the 20Ti stack. (e) Concentration of oxygen and titanium elements, and lattice-oxygen in the respective depth. (f) Conduction mechanism of the devices. Reprinted from [21].

the switching layer is limited to a great extent as can be seen by XPS depth spectra in **Figure 8(c)** and **(d)**. These results indicate towards the role of barrier layer TiW and interfacial layer TiWOx. The TiW insertion layer also promote confined filament which was not the case with device A having abundant Cu diffusion (**Figure 8e**). They suggest that the TiWOx interfacial layer promotes the formation of hybrid filament. Thus, device B shows superior device stability and performance compared to device A (pure metallic filament), as depicted in **Figure 8(e–f)**.

Nevertheless, Wan T. et al. [27] suggest that the digital to analog switching in Ag/$SrTiO_3$/FTO device can also be achieved with a pure metallic filament by controlling the size of the filament during the switching operation. They inserted a reduced graphene oxide (RGO) layer on top of the FTO bottom electrode. The RGO has high interfacial resistance and help to dissipate the Joule heat through the RGO. Hence, the size of the metallic filament can be easily tuned to perform good analog behavior.

2.3 Electrode engineering

Jang J.T. et al. [28] observed that a careful choice of top electrode material is crucial in achieving analog behavior. They compared Mo/IGZO/Pd (sample 1) and Pd/IGZO/Pd (sample 2) stacks. The devices made with the Pd top electrode exhibits digital switching, an abrupt transition in the resistance state during the set and reset operations are depicted in **Figure 9(a)**. On the other hand, the Mo/IGZO/Pd stack exhibits gradual transformation of resistance state for the set and reset operation, as shown in **Figure 9(b)**.

Energy band diagram analysis for sample 1 and sample 2 is depicted in **Figure 10(a)–(d)**. The abrupt switching in the Pd/IGZO/Pd stack is predominately due to a more significant barrier height of about 1 eV. A larger barrier height between the metal and semiconductor interface induces the formation of Schottky junction near the top and bottom electrode. However, the Mo/IGZO/Pd stack is observed to have a minimal barrier height of 0.3 eV, which significantly results in the formation of an ohmic junction near the top electrode and a Schottky junction near the bottom electrode. Thus, in the presence of an ohmic junction near the top electrode for sample 2, they observed a gradual switching behavior on engineering the top electrode. A similar phenomenon was also observed by Tang M.H. et al. [29] in Pt/ZnO/Pt and Ag/ZnO/Pt stacks devices. Digital unipolar switching behavior is observed in Pt/ZnO/Pt device and, conversely, the Ag/ZnO/Pt device exhibits analog bipolar switching. Although the paper does not discuss the detailed physics of such phenomenon, we assume that the contribution of Ag cations should play a role in achieving the analog behavior.

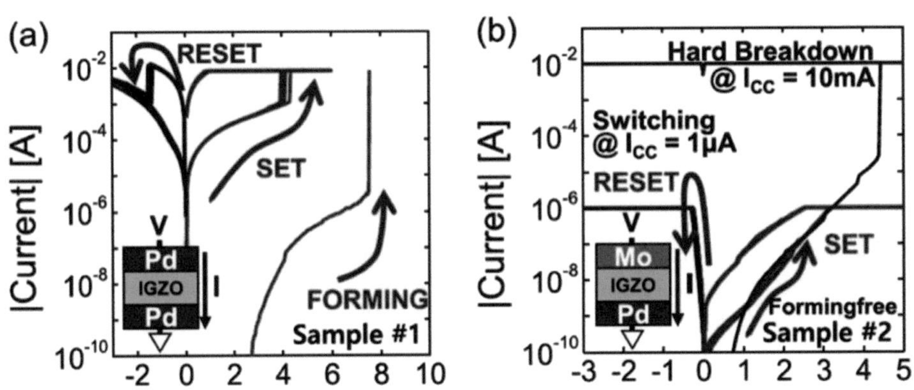

Figure 9.
Typical I-V curves of (a) Pd/IGZO/Pd (sample 1) and (b) Mo/IGZO/Pd (sample 2) reprinted from [28].

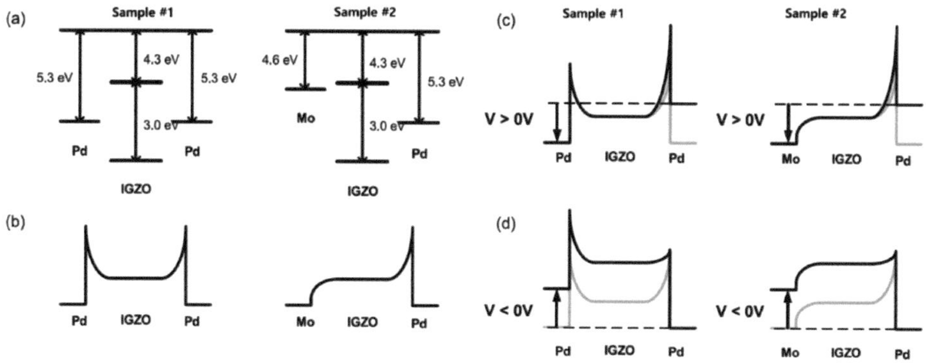

Figure 10.
The schematic of (a) flat band diagram, (b) equilibrium, (c) positive bias, and (d) negative bias energy band diagram of sample 1 and sample 2. Reprinted from [28].

Li X. et al. [30] investigated the role of inert and oxidizable top electrode materials in trilayer AlOx/TaOx/TaOy devices. The digital switching behavior is observed for the device with Pt as top electrode. The device made with Al top electrode tends to exhibit analog switching behavior. The Al top electrode interacts with the AlOx layer leading to the formation of oxygen deficient interfacial layer between AlOx and Al (top electrode). The oxygen ion migration and accumulation occur in a continuous manner and the device with Al top electrode is exhibiting gradual switching behavior during the continuous set and reset operation.

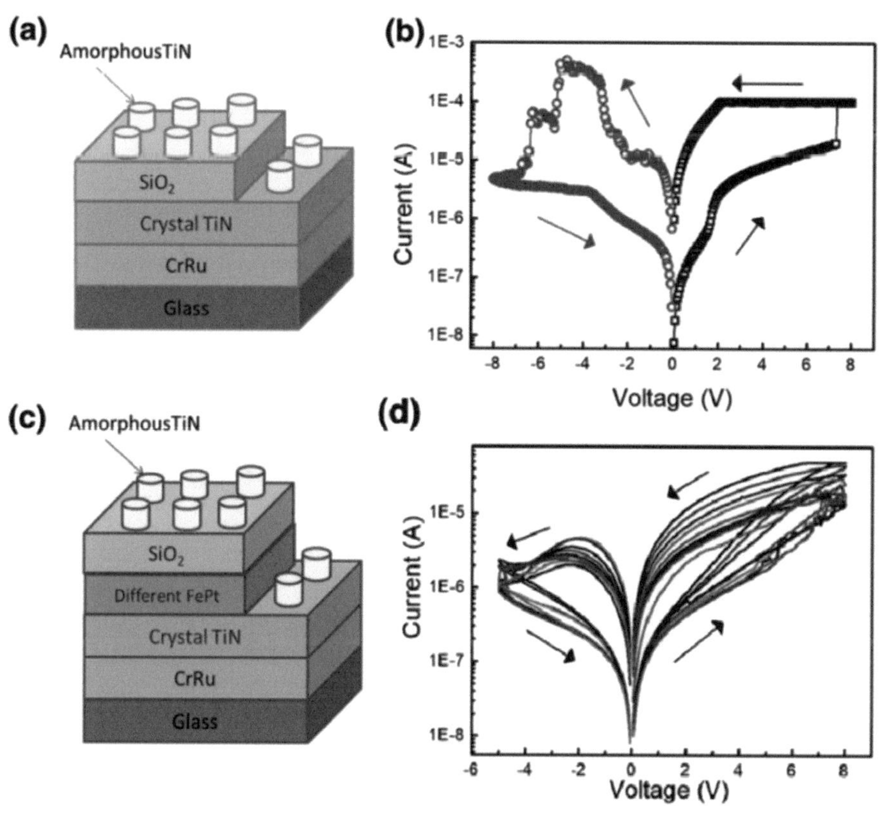

Figure 11.
The schematic and IV curve of (a) & (b) device a & (c) & (d) device B. reprinted from [31].

The method that Li X. et al. [30] proposed was further explored by C. Sun. et al. [31]. They suggest that an alloy of inert-oxidizable metal, FePt electrode, can induce analog characteristics on SiO2 based devices. The **Figure 11(a)** and **(c)** depict the schematic representation of Device A (TiN/SiO2/TiN) and Device B (TiN/SiO2/FePt/TiN). The FePt electrode act as the oxygen reservoir layer which assisting in trapping of oxygen ions during the resistive switching transformation. The barrier height of the electrode is also engineered on introducing FePt electrode which influence the switching transformation from abrupt to gradual as depicted in **Figure 11(b)** and **(d)**.

3. Conclusion

We overviewed some essential techniques to fabricate analog memristive devices. Based on the techniques that we covered here, we noticed that most of the analog memristors tend to have or work at a lower current or conduction than that of digital memristors; the resistance of the On and the Off states is higher prior to the digital-to-analog transformation. Henceforth, we can assume that the employment of low voltage or current favors the exhibition of analog behavior, and thus, the synaptic characteristics can be observed easier. We hypothesized that some other way that might work to induce digital-to-analog transformation in the memristors is to engineer the partial formation of the conductive filament; this can be done by intentionally decrease the compliance current and the voltage operation. **Table 1** summarizes the switching characteristics and synaptic behavior of several devices in reported studies. It is clear that the devices that endure digital-to-analog switching transformation often suffer from low memory window and dynamic

Device structure	Switching characteristics		Synaptic behavior		Note	References
	Analog/ digital	On/off ratio	Linearity	Dynamic range (%)		
Pt/ZnO/TiN	Analog	<10	NA	33	Bipolar switching	[10]
Pr/ZnO/TiN	Semi-analog	10	NA	12.5	Complementary switching, digital set	[10]
Cu/HfAlOx/Si	Digital	10^6	NA	125	Full reset, abrupt potentiation and depression	[10]
Cu/HfAlOx/Si	Analog	10	NA	12	Partial reset	[20]
TiN/TiO_2/TiN	Digital	>10	NA	800	Abrupt depression	[21]
TiN/TiO_2/TiN	Analog	±10	NA	26	Multibit	[21]
Cu/TaOx/TiN	Semi-analog	Unstable	0.87	150	Digital set, abrupt potentiation	[26]
Cu/TiW/TaOx/ TiN	Analog	>10	0.69	34	Multibit	[26]

From Simanjuntak et al.

Table 1.
Several important examples of digital-to-analog switching transformation in published literature.

range. We assume that analog switching in memristor device can only work when the switching region in the cell is small or no major structural changes happen due to the repeated switching cycles or pulses.

We also addressed several challenges in the analog memristor devices that could hinder the commercialization of these devices. First, the non-linear switching modes, such as complementary and diode-like modes, have unique advantages in integrating the memristors in the array configurations; however, the synaptic reliability of the devices having these modes is still less understood, and extensive investigation is needed to study the efficacy of these modes on the synaptic plasticity. Secondly, analog memristors often suffer from short retention [14], which may not be suitable for data storage applications; however, this property opens another opportunity to modulate the long- and short-term memory plasticity of the devices that can be useful for neuromorphic computing applications. Hence, the trade-off between short retention and synaptic plasticity should be carefully managed to fit with the desired application. Third, the cycle-to-cycle operation can promote internal cell resistance variability that degenerates the analog behavior and could lead to the occurrence of digital switching [32]. Analog devices made with interfacial layer techniques may suffer from this problem due to the repeated redox reactions at the interfacial layer. Therefore, we hypothesized that a careful choice of the insertion layer plays a crucial role in achieving long cycle endurance and, thus, may prolong the exhibition of analog behavior.

Acknowledgements

This work was supported by the EPSRC Programme Grant (No. EP/R024642/1) and the H2020- FETPROACT-2018-01 SYNCH project. D. Panda thanks to DST-SERB research grant, Government of India (Grant No. SRG/2019/000129) to support this work.

Conflict of interest

The authors declare no conflict of interest.

Author details

Firman Mangasa Simanjuntak[1*], Sridhar Chandrasekaran[2], Debashis Panda[3], Aftab Saleem[4] and Themis Prodromakis[1]

1 Centre for Electronics Frontiers, University of Southampton, Southampton, United Kingdom

2 Advanced Research Institute, Dr. MGR Educational and Research Institute, Chennai, India

3 Department of Physics, National Institute of Science and Technology, Odisha, India

4 Department of Electrical Engineering and Computer Science, National Yang Ming Chiao Tung University, Hsinchu, Taiwan

*Address all correspondence to: f.m.simanjuntak@soton.ac.uk

References

[1] Salahuddin S, Ni K, Datta S. The era of hyper-scaling in electronics. Nat. Electron. 2018;1:442-450. DOI:10.1038/s41928-018-0117-x.

[2] Xi Y, Gao B, Tang J, Chen A, Chang M-F, Hu XS, et al. In-memory Learning with Analog Resistive Switching Memory: A Review and Perspective. Proc. IEEE 2021;109:14-42. DOI:10.1109/JPROC.2020.3004543.

[3] Li C, Belkin D, Li Y, Yan P, Hu M, Ge N, et al. Efficient and self-adaptive in-situ learning in multilayer memristor neural networks. Nat. Commun. 2018;9:7-14. DOI:10.1038/s41467-018-04484-2.

[4] Zhao M, Gao B, Tang J, Qian H, Wu H. Reliability of analog resistive switching memory for neuromorphic computing. Appl. Phys. Rev. 2020;7:011301. DOI:10.1063/1.5124915.

[5] Stathopoulos S, Khiat A, Trapatseli M, Cortese S, Serb A, Valov I, et al. Multibit memory operation of metal-oxide bi-layer memristors. Sci. Rep. 2017;7:17532. DOI:10.1038/s41598-017-17785-1.

[6] Simanjuntak FM, Ohno T, Chandrasekaran S, Tseng T-Y, Samukawa S. Neutral oxygen irradiation enhanced forming-less ZnO-based transparent analog memristor devices for neuromorphic computing applications. Nanotechnology 2020;31:26LT01. DOI:10.1088/1361-6528/ab7fcf.

[7] Chandrasekaran S, Simanjuntak F, Saminathan R, Panda D, Tseng T-Y. Improving linearity by introducing Al in HfO2 as memristor synapse device. Nanotechnology 2019;6:107-113. DOI:10.1088/1361-6528/ab3480.

[8] Simanjuntak FM, Ohno T, Samukawa S. Film-Nanostructure-Controlled Inerasable-to-Erasable Switching Transition in ZnO-Based Transparent Memristor Devices: Sputtering-Pressure Dependency. ACS Appl. Electron. Mater. 2019;1:2184-2189. DOI:10.1021/acsaelm.9b00617.

[9] Celano U, Goux L, Belmonte A, Giammaria G, Opsomer K, Detavernier C, et al. Progressive vs. abrupt reset behavior in conductive bridging devices: A C-AFM tomography study. 2014 IEEE Int. Electron Devices Meet., IEEE; 2014, p. 14.1.1-14.1.4. DOI:10.1109/IEDM.2014.7047048.

[10] Khan SA, Lee GH, Mahata C, Ismail M, Kim H, Kim S. Bipolar and Complementary Resistive Switching Characteristics and Neuromorphic System Simulation in a Pt/ZnO/TiN Synaptic Device. Nanomaterials 2021;11:315. DOI:10.3390/nano11020315.

[11] Rosezin R, Linn E, Kugeler C, Bruchhaus R, Waser R. Crossbar Logic Using Bipolar and Complementary Resistive Switches. IEEE Electron Device Lett. 2011;32:710-712. DOI:10.1109/LED.2011.2127439.

[12] Lee J, Shin J, Lee D, Lee W, Jung S, Jo M, et al. Diode-less nano-scale ZrOx/HfOx RRAM device with excellent switching uniformity and reliability for high-density cross-point memory applications. Tech. Dig. - Int. Electron Devices Meet. IEDM 2010:452-455. DOI:10.1109/IEDM.2010.5703393.

[13] Chen J, Hsin C, Huang C, Chiu C, Huang Y. Dynamic Evolution of Conducting Nano filament in Resistive Switching Memories. Nano Lett. 2013.

[14] Huang C-H, Huang J-S, Lai C-C, Huang H-W, Lin S-J, Chueh Y-L. Manipulated transformation of filamentary and homogeneous resistive switching on ZnO thin film memristor with controllable multistate. ACS Appl.

Mater. Interfaces 2013;5:6017-6023. DOI:10.1021/am4007287.

[15] Ryu H, Kim S. Synaptic Characteristics from Homogeneous Resistive Switching in Pt/Al2O3/TiN Stack. Nanomaterials 2020;10:2055. DOI:10.3390/nano10102055.

[16] Biju KP, Liu X, Kim S, Jung S, Park J, Hwang H. Coexistence of filamentary and homogeneous resistive switching in graded WOx thin films. Phys. Status Solidi - Rapid Res. Lett. 2011;5:89-91. DOI:10.1002/pssr.201004455.

[17] Li Y, Chu J, Duan W, Cai G, Fan X, Wang X, et al. Analog and Digital Bipolar Resistive Switching in Solution-Combustion-Processed NiO Memristor. ACS Appl. Mater. Interfaces 2018;10: 24598-24606. DOI:10.1021/acsami.8b05749.

[18] Yi H, Ilyas N, Li C, Li D, Jiang X, Li W. Coexistence of Digital and Analog Resistive Switching Behaviours in Ag/ CuAlO 2 /TiO 2 /p ++ -Si Memristor. J. Phys. Conf. Ser. 2020;1637:012053. DOI:10.1088/1742-6596/1637/1/012053.

[19] Muenstermann R, Menke T, Dittmann R, Waser R. Coexistence of Filamentary and Homogeneous Resistive Switching in Fe-Doped SrTiO3 Thin-Film Memristive Devices. Adv. Mater. 2010;22:4819-4822. DOI:10.1002/ adma.201001872.

[20] Kim S, Kwon O, Ryu H, Kim S. Improved Synaptic Device Properties of HfAlOx Dielectric on Highly Doped Silicon Substrate by Partial Reset Process. Metals (Basel). 2021;11:772. DOI:10.3390/met11050772.

[21] Chang L-Y, Simanjuntak FM, Hsu C-L, Chandrasekaran S, Tseng T-Y. Suboxide interface induced digital-to-analog switching transformation in all Ti-based memristor devices. Appl. Phys. Lett. 2020;117:073504. DOI:10.1063/5.0014829.

[22] Simanjuntak FM, Chandra sekaran S, Lin C-C, Tseng T-Y. Switching Failure Mechanism in Zinc Peroxide-Based Programmable Metallization Cell. Nanoscale Res. Lett. 2018;13:327. DOI:10.1186/s11671-018-2743-7.

[23] Simanjuntak FM, Ohno T, Samukawa S. Neutral Oxygen Beam Treated ZnO-Based Resistive Switching Memory Device. ACS Appl. Electron. Mater. 2019;1:18-24. DOI:10.1021/ acsaelm.8b00055.

[24] Panda D, Simanjuntak FM, Chandrasekaran S, Pattanayak B, Singh P, Tseng T-Y. Barrier Layer Induced Switching Stability in Ga:ZnO Nanorods Based Electrochemical Metallization Memory. IEEE Trans. Nanotechnol. 2020;19:764-768. DOI:10.1109/TNANO.2020.3029588.

[25] Chandrasekaran S, Simanjuntak FM, Tseng T. Controlled resistive switching characteristics of ZrO 2 -based electrochemical metallization memory devices by modifying the thickness of the metal barrier layer. Jpn. J. Appl. Phys. 2018;57:04FE10. DOI:10.7567/ JJAP.57.04FE10.

[26] Saleem A, Simanjuntak FM, Chandrasekaran S, Rajasekaran S, Tseng T-Y, Prodromakis T. Transformation of digital to analog switching in TaOx-based memristor device for neuromorphic applications. Appl. Phys. Lett. 2021;118:112103. DOI:10.1063/5.0041808.

[27] Wan T, Qu B, Du H, Lin X, Lin Q, Wang DW, et al. Digital to analog resistive switching transition induced by graphene buffer layer in strontium titanate based devices. J. Colloid Interface Sci. 2018;512:767-774. DOI:10.1016/j.jcis.2017.10.113.

[28] Jang JT, Min J, Hwang Y, Choi SJ, Kim DM, Kim H, et al. Digital and analog switching characteristics of

InGaZnO memristor depending on top electrode material for neuromorphic system. IEEE Access 2020;8:192304-192311. DOI:10.1109/ACCESS.2020.3032188.

[29] Tang MH, Jiang B, Xiao YG, Zeng ZQ, Wang ZP, Li JC, et al. Top electrode-dependent resistance switching behaviors of ZnO thin films deposited on Pt/Ti/SiO2/Si substrate. Microelectron. Eng. 2012;93:35-38. DOI:10.1016/j.mee.2011.12.003.

[30] Li X, Wu H, Bin Gao, Wu W, Wu D, Deng N, et al. Electrode-induced digital-to-analog resistive switching in TaO x -based RRAM devices. Nanotechnology 2016;27:305201. DOI:10.1088/0957-4484/27/30/305201.

[31] Sun C, Lu SM, Jin F, Mo WQ, Song JL, Dong KF. Investigation of FePt electrode induced influence on resistive switching characteristics of SiO2-based RRAM. J. Mater. Sci. Mater. Electron. 2020;31:19989-19996. DOI:10.1007/s10854-020-04521-1.

[32] Hardtdegen A, La Torre C, Zhang H, Funck C, Menzel S, Waser R, et al. Internal Cell Resistance as the Origin of Abrupt Reset Behavior in HfO2-Based Devices Determined from Current Compliance Series. 2016 IEEE 8th Int. Mem. Work. IMW 2016 2016:2-5. DOI:10.1109/IMW.2016.7495280.